AIRCRAFT PROPELLERS AND CONTROLS

By Frank Delp

International Standard Book Number 0-89100-097-6
For sale by: IAP, Inc., A Hawks Industries Company
Mail To: P.O. Box 10000, Casper, WY 82602-1000
Ship To: 7383 6WN Road, Casper, WY 82604-1835
(800) 443-9250 ❖ (307) 266-3838 ❖ FAX: (307) 472-5106

IAP, Inc.
7383 6WN Road, Casper, WY 82604-1835

Printed in the United States of America

Table Of Contents

Preface

This book on *Aircraft Propellers and Controls* is one of a series of specialized training manuals prepared for aviation maintenance personnel.

This series is part of a programmed learning course developed and produced by International Aviation Publishers (IAP), one of the largest suppliers of aviation maintenance training materials in the world. This program is part of a continuing effort to improve the quality of education for aviation mechanics throughout the world.

This manual is designed to present the A&P mechanic/student with the information necessary for a general understanding of the theory, operation and maintenance of fixed-pitch, variable-pitch, feathering, and reversing propellers. Although specific propeller systems are discussed in some chapters, the systems operate in a manner generally applicable to all similar propeller systems.

The information contained in this manual is for instructional purposes only, and is not to be used as a substitute for a manufacturers current maintenance manual, service bulletins or operational data.

Throughout this text, at appropriate points, is included a series of carefully prepared questions and answers to emphasize key elements of the text, and to encourage the individual to continually test himself for accuracy and retention as he progresses. A multiple choice final examination is included to allow you to test your comprehension of the total material.

Some of the terminology used in this book may be new to you. Throughout the text, you will find terms that are defined in the Glossary at the back of the book highlighted as follows: ***glossary item.***

If you have any questions or comments regarding this manual, or any of the many other textbooks offered by IAP, simply contact: Sales Department, IAP, Inc.; Mailing Address: P.O. Box 10000, Casper, WY 82602-1000; Shipping Address: 7383 6WN Road, Casper, WY 82604-1835; or call toll free: (800) 443-9250; International, call: (307) 266-3838.

Chapter I
Introduction To Propellers

Throughout the development of controlled flight as we know it, every aircraft required some kind of device to convert engine power to a usable form termed *thrust*. With few exceptions, nearly all of the early practical aircraft designs used ***propellers*** to create necessary thrust. During the latter part of the 19th century many unusual and innovative designs for propellers made their debut on the early flying machines. These ranged from simple wood frame and fabric paddles to elaborate multi-bladed wire-braced designs. Some of these designs were even used successfully as a means of propelling the early dirigibles and heavier-than-air designs.

As the science of aeronautics progressed, propeller designs improved from flat boards which merely pushed air backward, to actual airfoil shapes that produced *lift*, as do wings, to pull the aircraft forward by aerodynamic action. By the time the Wright brothers began their first powered flights, propeller designs had evolved into the standard two-bladed style similar in appearance to those used on today's modern light aircraft.

World War I brought about an increase in aircraft size, speed, and engine horsepower requiring further improvements in propeller designs. The most widely used improvement during the war was a *four-bladed* wood propeller. Other design improvements, which were developed during the war, included an aluminum ***fixed-pitch propeller***, and the ***two-position controllable propeller***. These improvements did not come into wide usage until the late 1920s.

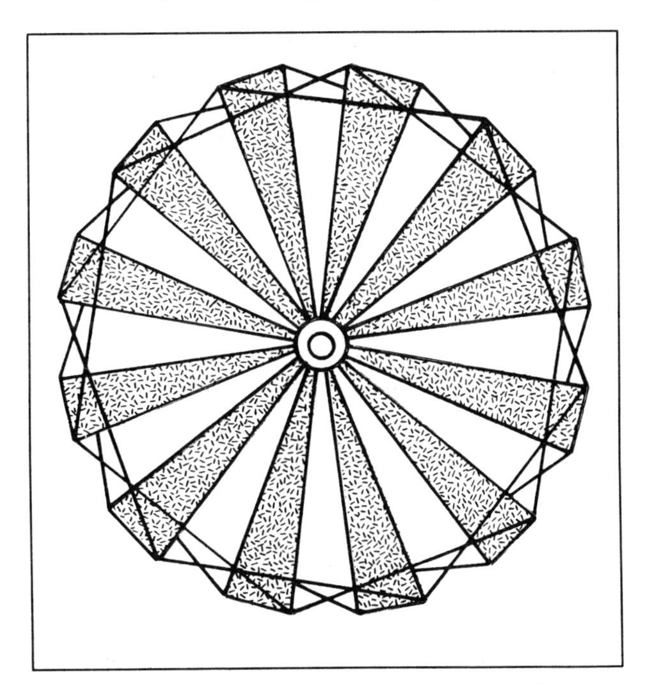

Figure 1-1. An 1874 propeller design from the first powered airplane to leave the ground.

As aircraft designs improved, propellers were developed which featured thinner airfoil sections, and greater strength. Because of its structural strength, these improvements brought the aluminum propeller into wide usage. The advantage of being able to change the propeller ***blade angle*** in flight, led to wide acceptance of the two-position propeller, and later, the development of the ***constant-speed propeller system.*** This same constant-speed propeller system is still in use on many propeller driven aircraft being produced today.

Refinements of propeller designs and systems from the 1930s through World War II included the ***featherable*** propeller for multi-engine applications, ***reversing*** propeller systems which allowed for shorter landing runs and improved ground maneuverability, and many special auxiliary systems such as ice elimination, simultaneous control systems, and automatic feathering systems. Moreover, with the development of the jet engine, propeller systems were adapted for use with these engines to allow their efficient use at low altitudes and low airspeeds.

Today, propeller designs continue to be improved by the use of new airfoil shapes, composite materials, and multi-blade configurations. Recent improvements include the use of laminar and symmetrical airfoils, fiberglass materials, and gull wing propeller designs.

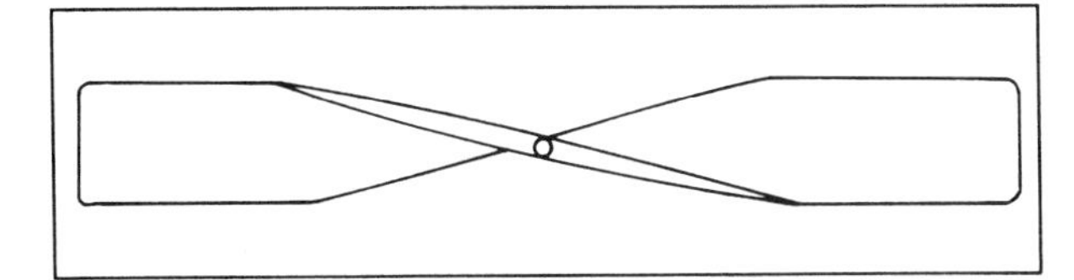

Figure 1-2. Propeller used by the Wright Brothers in 1911.

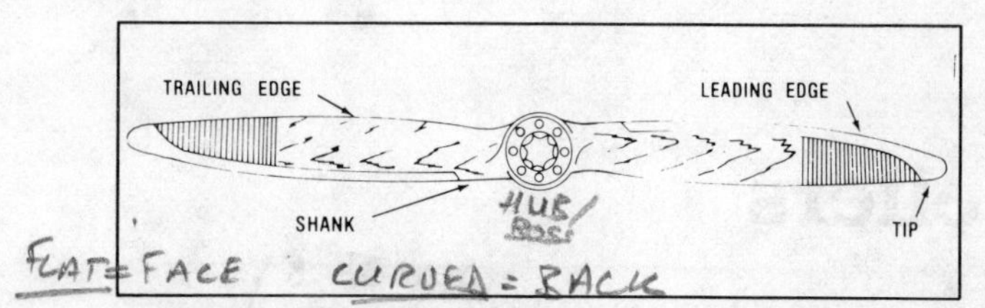

Figure 1-3. Parts of a propeller.

A. Nomenclature

Before starting any discussion about propellers, it is necessary that some basic terms be defined to avoid confusion and misunderstanding.

First of all, exactly what is a propeller? A propeller normally consists of two or more ***blades*** attached to a central ***hub*** which is mounted on an engine crankshaft. The purpose of the propeller is to convert engine horsepower to useful thrust. The blades, which are actually rotating wings, have a ***leading edge***, trailing edge, ***tip, shank, face*** and ***back*** as shown in Figures 1-3 and 1-4. Many people have trouble with the terms *face* and *back* so it is helpful to visualize the flat side of the blade as facing the pilot when he is in the cockpit.

A term that will be used throughout this text is *blade angle*. This is the angle between the propeller ***plane of rotation*** and the ***chord line*** of a propeller airfoil section. Another term, ***blade station*** is a reference position on a blade that is a specified distance from the center of the hub. ***Pitch*** is the distance (in inches) that a propeller section will move forward in one revolution. ***Pitch distribution*** is the gradual twist in the propeller blade from shank to tip.

Editor's Note: A complete listing of general propeller terms can be found in the Glossary at the end of this book.

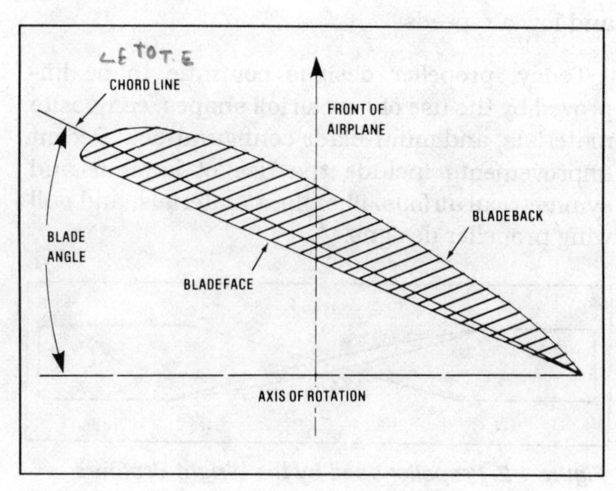

Figure 1-4. Propeller blade cross section.

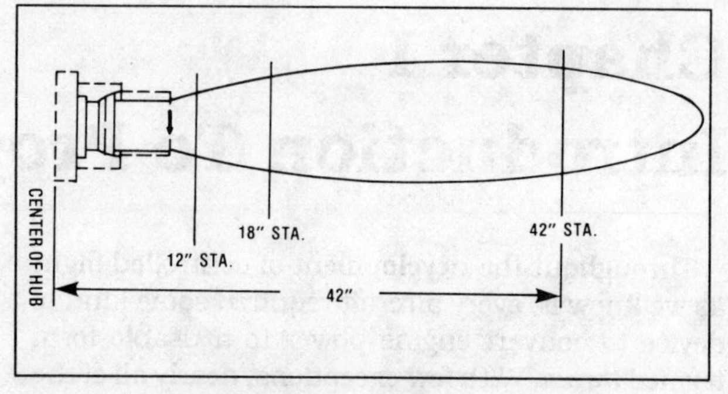

Figure 1-5. Propeller blade stations

B. Types Of Propellers

Since the beginning of powered flight, many unique designs for propellers have appeared with only a few becoming widely accepted. The following are brief descriptions of the more common designs that are being used in aviation.

The *fixed-pitch* propeller is the most widely used propeller design in aviation. A fixed-pitch propeller may be made of wood, aluminum, or steel and is considered to be of one-piece construction with a blade angle that cannot normally be changed. Fixed-pitch propellers are usually found on light single-engine aircraft.

Ground-adjustable propellers are similar to fixed-pitch propellers in that their blade angles cannot be changed in flight. However, since the propeller is designed so that the blade angles can be changed on the ground, the propeller can be adjusted to give the desired propeller characteristics for a flight (i.e., low blade angle for taking off from a short field or high blade angle for more speed on a cross-country flight). This type of propeller was widely used on aircraft built in the 1920s, '30s, and '40s.

The *two-position* propeller is a design that allows the pilot to select one of two blade angles while in flight, allowing the use of a low blade angle for takeoff and a high blade angle for cruise. This is something like a two-speed transmission in an automobile. The two-position propeller was used on some of the more sophisticated designs in the late 1920s and '30s.

Controllable-pitch propellers were designed to give the pilot the ability to set the blades at any angle in the propeller's range while in flight, giving the pilot more control over the propeller than the two-position design. This style of propeller was popular on light aircraft in the 1940s due to its low weight and simple mechanism.

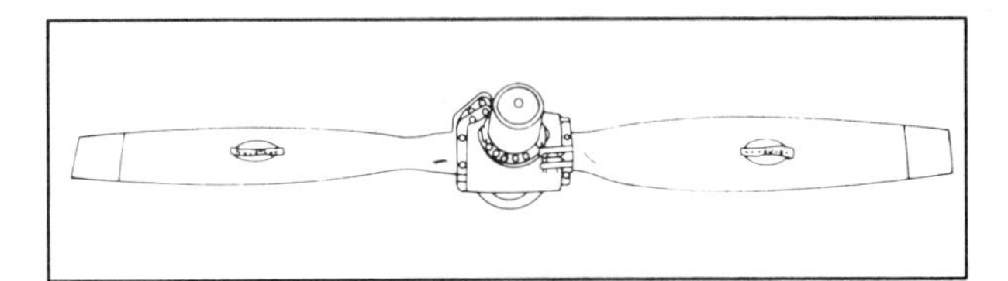

Figure 1-6. Modern constant-speed feathering propeller.

The most *imaginative* propeller design to come into existence has been the *automatic pitch-changing* propeller. This propeller is not controllable by the pilot. Instead, it will theoretically set the propeller blades at the most efficient angle by reacting to forces generated by engine thrust torque, and airspeed. This style propeller was used on aircraft produced between the late 1930s and the 1950s. While designed for engines up to 450 hp, most aircraft using these propellers today have engines of 150 hp or less.

Most medium and high performance aircraft produced today are equipped with *constant-speed* propeller systems. This propeller system uses a controllable propeller which the pilot indirectly controls by adjusting a constant-speed control unit, commonly called the ***governor***. Propeller blade angle is adjusted by this governor to maintain the engine speed (RPM) which the pilot has set on the governor. Due to this controllability, coupled with the relatively lightweight and low cost of modern constant speed systems, some earlier propeller designs have become less common.

Most multi-engine aircraft equipped with constant-speed propeller systems also have the capability of feathering the propeller. When a propeller is feathered, the propeller blades are rotated to present an edge to the wind, eliminating the drag associated with a windmilling propeller when an engine fails.

Reversing propeller systems are refinements of the constant-speed feathering systems. The propeller blades can be rotated to a *negative* angle to create negative thrust. This forces air forward instead of backwards permitting a shorter landing roll and improved ground maneuvering. Reversing propeller systems are usually found on the more sophisticated multi-engine aircraft.

C. The Propeller Protractor

A useful propeller tool is the *universal propeller protractor*. This device is used to measure the propeller blade angle at a specific blade station to determine if the propeller is properly adjusted. The blade angle is referenced from the propeller plane of rotation, which is ninety degrees to the crankshaft centerline.

The frame of the protractor is made of aluminum and has three sides which are ninety degrees to each other. A level is mounted on one corner of the front of the frame. This corner spirit level swings out and is used to indicate when the protractor is vertical. A movable ring is located in the frame and is used to set the *zero reference angle* for blade angle measurements. The ring is engraved with index marks which allow readings as small as one-tenth of a degree. A center disc is engraved with a degree scale from 0 to 180 degrees positive and negative and contains a spirit level to indicate when the disc is level. Locking and adjusting controls are shown in Figure 1-7.

When using this device and before measuring the angle of a propeller blade, the *reference blade station* must be determined from the propeller or aircraft manufacturer's maintenance manual. This station will normally be 30-inch, 36-inch, or 42-inch. The reference station should be marked with chalk or a grease pencil on the face of each blade. **CAUTION:** *Do not use a graphite (black lead) pencil as it will cause corrosion!*

The next procedure is to establish the *reference plane* from the engine crankshaft centerline. The reference plane is not based on the airframe attitude because of the canted installation of some engines.

To zero the protractor you must loosen the ring-to-frame lock, align the zeros on the disc and the ring, and then engage the disc-to-ring lock. Place one edge of the protractor on a flat propeller hub surface that is parallel or perpendicular to the crankshaft centerline. Turn the ring adjustor until the spirit level in the center of the disc is level. (The corner spirit level should also be level.) Now, tighten the ring-to-frame lock and release the disc-to-ring lock. The protractor is now aligned with the engine centerline.

Place one blade of the propeller horizontal and move out to the reference station marked on the face of the blade to measure the angle. Stand on the same side of the airplane, facing in the same direction as when establishing the zero with the protractor, otherwise the measurements will be incorrect. Place the edge of the protractor on the face at the reference station and turn the disc adjuster until the spirit level centers and read the blade angle using the zero line on the ring as the index.

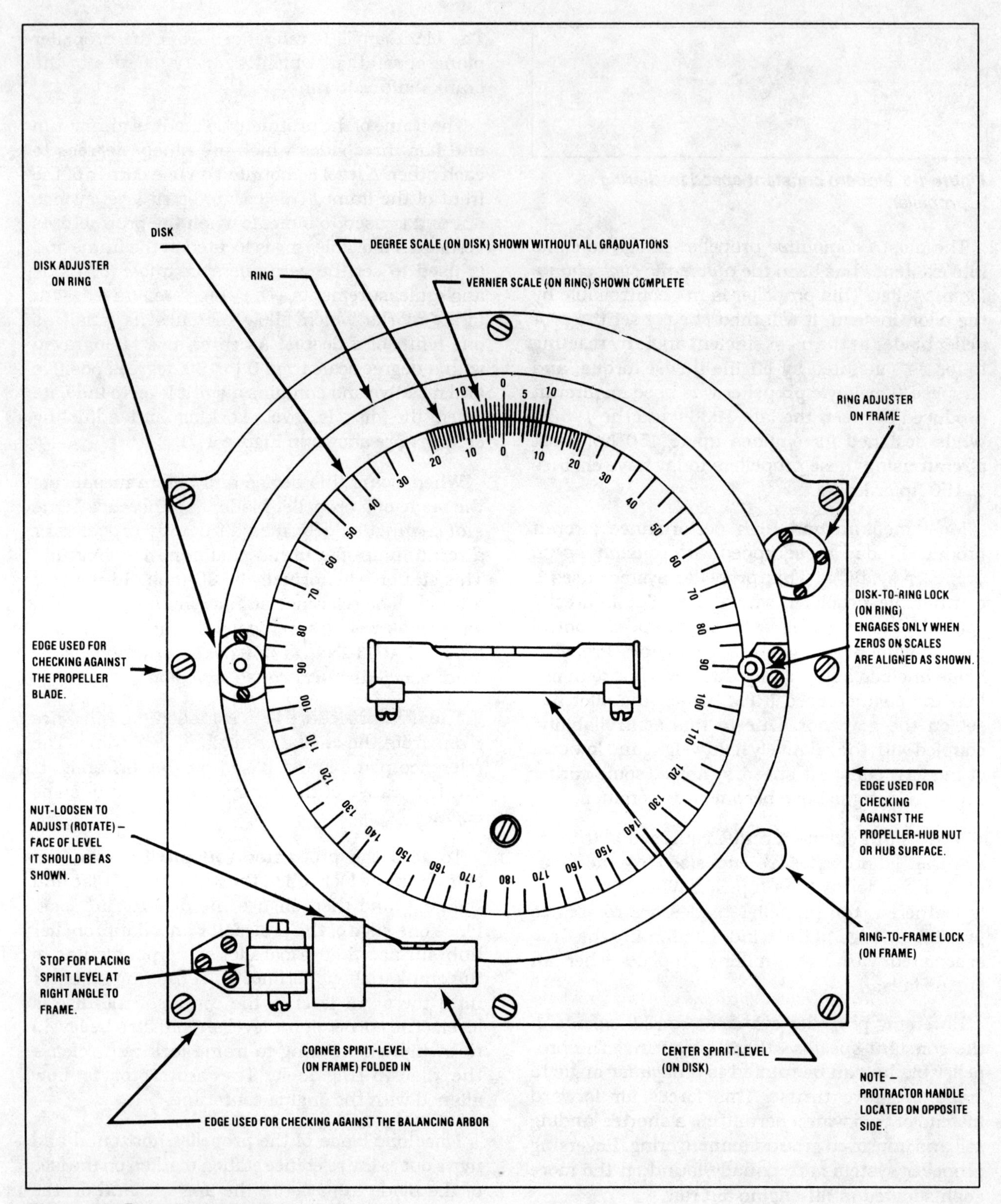

Figure 1-7. Propeller protractor.

Tenths of degrees can be read from the vernier scale. Rotate each blade to the same horizontal position and measure the angle. The amount of allowed angle variation among the blades will vary with each design.

If the face of the blade is curved, use masking tape to attach a piece of 1/8-inch rod (drill bits will do) 1/2-inch in from the leading and trailing edge and measure the angle with the protractor resting on the rods.

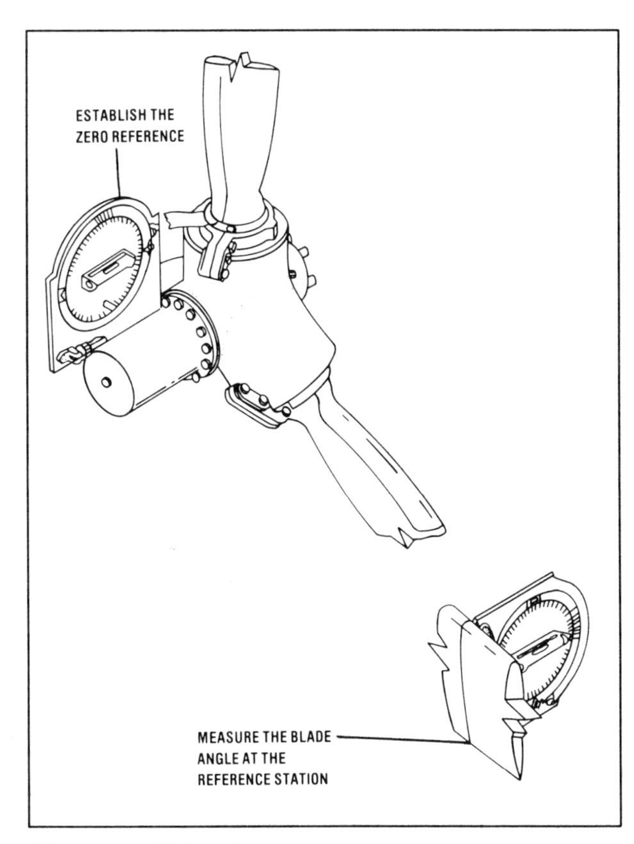

Figure 1-8. Using the protractor.

D. Propeller Safety

The aircraft propeller is the deadliest component on an aircraft. Obviously, when the engine is running, it is not advisable to put anything in the propeller arc, but how about when the engine is not running? Many people have been injured while *pulling an engine through* by hand. The engine fired with disastrous results! Remember, it is not possible to move a propeller on an engine without the possibility of the engine firing. **Just because the magneto switch is "Off" does not mean that the system is safe!** P-leads can break, magnetos can be defective, hot carbon may be in a cylinder immediately after engine shutdown; combine this with the proper fuel-air mixture in the cylinder and the engine can fire.

Before working on a propeller installation make sure that the magneto switch is off and if the switch incorporates a key, remove it. Place the mixture control in the idle cutoff position, and see that the aircraft master switch is in the off position. It is a good practice to mark these items with a red tag while maintenance is being performed. The only sure way, however, to prevent the engine from firing is to remove all spark plug leads from the spark plugs and remove one spark plug from each cylinder.

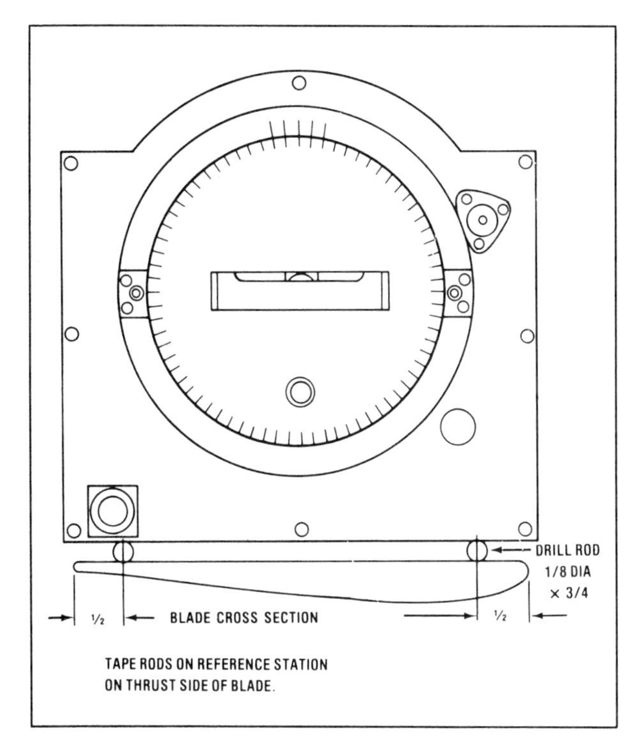

Figure 1-9. Allowing for blade curvature.

If you must work on an operable engine, *stay out of the arc of the propeller* and, unless absolutely necessary, *do not move the propeller!* Also, unless necessary for the work being performed, *do not allow anyone in the cockpit while you are working around the propeller!* They may turn on the magneto switch or engage the starter. It has happened!

QUESTIONS:

1. *What is the purpose of the propeller?*
2. *Which area of the propeller blade is referred to as the back?*
3. *Define blade angle.*
4. *What is the term given to the gradual twist of a propeller blade from shank to tip?*
5. *What is a ground-adjustable propeller?*
6. *What is a major advantage associated with reversing propellers?*
7. *How many spirit levels are centered when adjusting the propeller protractor?*
8. *What is the smallest division that can be read with a propeller protractor?*
9. *What should be the condition of the cockpit controls before working near a propeller?*
10. *Can the magneto fire if the magneto switch is turned off?*

Chapter II
FARs And Propellers

To understand the guidelines set down by the FAA regarding propeller system designs and maintenance, it is necessary to understand some of the regulations concerning propellers. This will be accomplished by looking at parts of the following Federal Aviation Regulations:

FAR Part 23, *Airworthiness Standards: Normal, Utility, and Acrobatic Aircraft*, and FAR Part 25, *Airworthiness Standards: Transport Category Aircraft*, outline the requirements for propellers and their control systems for aircraft certification. Because there is very little difference in the wording of Parts 23 and 25, they are considered identical for the purposes of this discussion.

Part 43 of the FAR defines the different classes of maintenance for the propeller system and the minimum requirements for 100-hour and annual inspections.

The information that is required to be permanently affixed to a propeller is discussed briefly with references to FAR Part 45, *Identification and Registration Markings*.

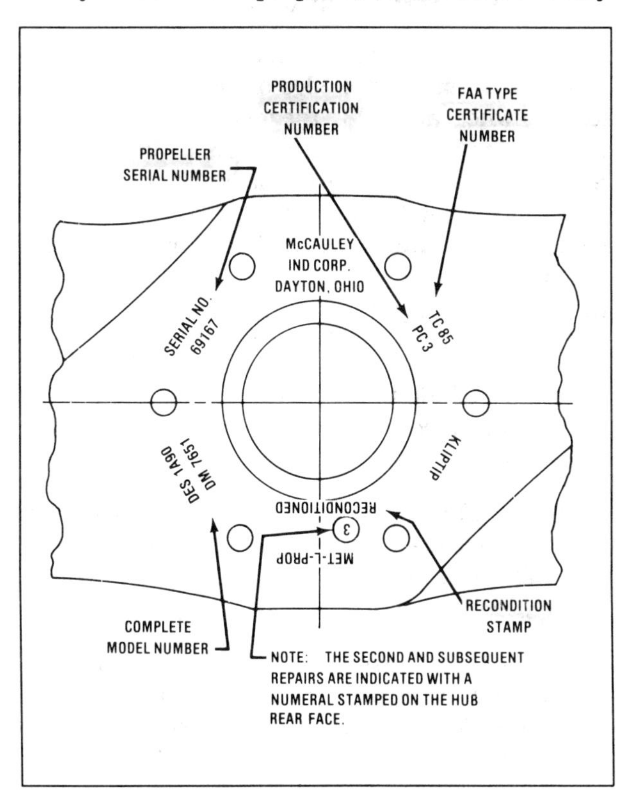

Figure 2-1. Typical hub stamping on a fixed-pitch propeller.

The licenses required to perform or supervise the maintenance or repair of a propeller and related systems are covered by FAR Part 65, *Certification: Airmen Other Than Flight Crewmembers*. This section distinguishes between the authority of a powerplant mechanic, a propeller repairman, and an authorized inspector.

Subparts of each FAR have been arranged for ease of presentation and do not follow the order as written in the FARs. The applicable regulations are noted at the appropriate points in this text.

A. Propeller Requirements For Aircraft Certification

Aircraft propellers must be certificated under FAR Part 35 and must contain the following information on the hub or butt of the propeller blade: builder's name, model designation, serial number, type certificate number, and production certificate number (FAR 45.13).

1. Static RPM

An aircraft which uses a fixed-pitch propeller will not operate at maximum RPM (tachometer redline) on the ground in a no-wind condition when the engine is producing maximum allowable horsepower. This is designed into the system to satisfy the requirement that the propeller must limit engine RPM to the maximum allowable when the engine is operating at full power and the aircraft is flying at its best rate-of-climb speed, thus preventing engine damage due to overspeeding. The propeller must also prevent the engine from exceeding the rated RPM by no more than 10% in a closed throttle dive at the aircraft's never-exceed speed (FAR 23/25.33). As airspeed or wind speed increases, engine RPM will increase because it is easier for the propeller to rotate. This explains why the listed ***static RPM*** (RPM at full power, on the ground, with no wind) for an aircraft is less than the engine rated RPM (tachometer redline).

All two-position and controllable-pitch propellers must comply with FAR 23/25.33 at their low blade angle setting.

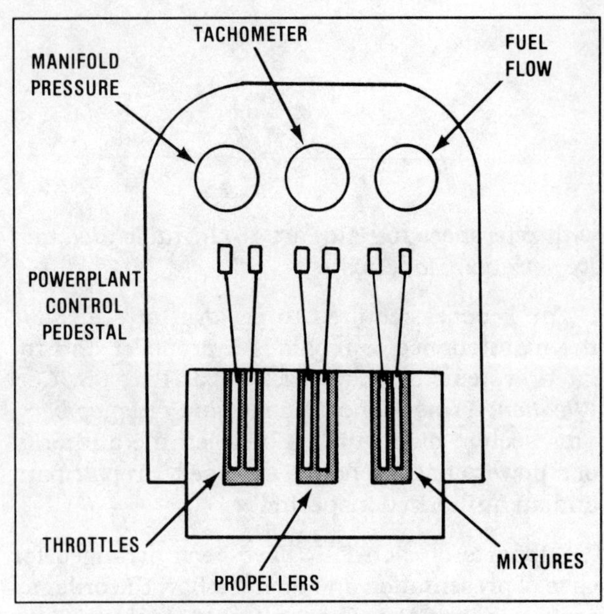

Figure 2-2. A typical light twin powerplant controls arrangement.

A constant-speed propeller system must limit engine speed to rated RPM at all times when the system is operating normally. If the governor should fail, the system must be designed to prevent a static RPM of no more that 103% of rated RPM (FAR 23/25.33). This is accomplished by using the correct low blade angle setting, the greater the blade angle, the lower the static RPM.

2. Cockpit Controls And Instruments

Propeller control levers in the cockpit must be arranged to allow easy operation of all controls at the same time, but not to restrict the movement of individual controls (FAR 23/25.1149).

The propeller controls must be rigged so that an increase in RPM is achieved by moving the controls forward and a decrease in RPM is caused by moving the controls aft. The throttles must be arranged so that forward thrust is increased by forward movement of the control and reverse thrust is increased by aft movement of the throttle (FAR 23/25.779). (When operating in reverse, the throttles are used to place the propeller blades at a negative angle.)

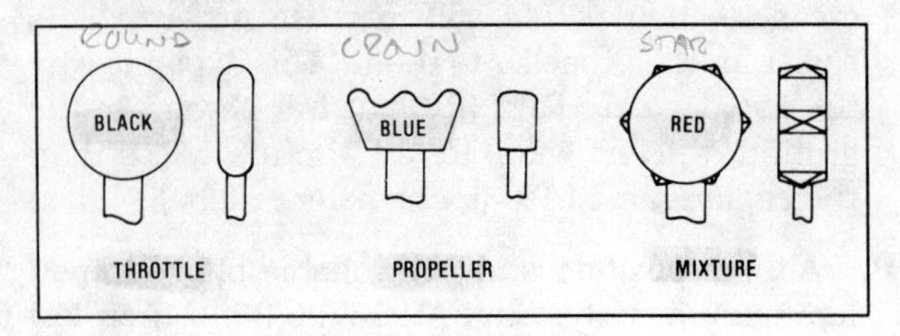

Figure 2-3. Powerplant control knobs.

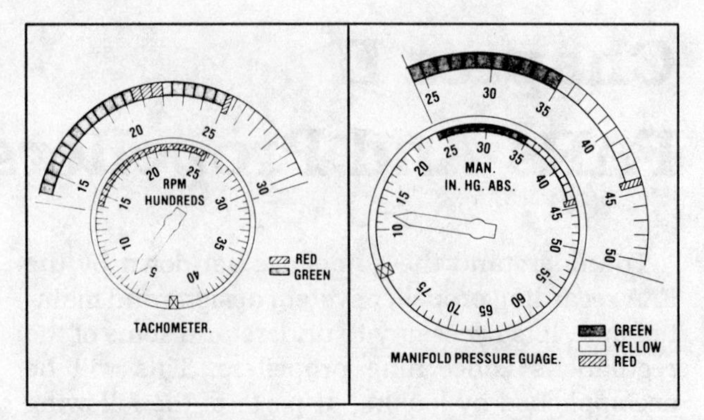

Figure 2-4. Tachometer and manifold pressure gauges and markings.

Cockpit powerplant controls must be arranged to prevent confusion as to which engine they control. Recent regulation changes require that control knobs be distinguished by shape and color (FAR 23/25.781) as shown in Figure 2-3.

Cockpit instruments such as tachometers and manifold pressure gauges must be marked with a *green arc* to indicate the normal operating range, a *yellow arc* for takeoff and precautionary range, a *red arc* for critical vibration range, and a *red radial line* for maximum operating limit (FAR 23/25.1549).

3. Minimum Terrain And Structural Clearances

The minimum ground clearance (distance from the level ground to the edge of the ***propeller disc***) for a tailwheel aircraft in the takeoff attitude is nine inches. For a tricycle-geared aircraft in the most nose-low normal attitude (stationary, taxi, or takeoff attitude) the minimum ground clearance is seven inches. These clearances are based on normal tire and strut inflation. If the tire and strut are deflated there need be only a positive ground clearance (the propeller disc must not touch the ground).

For a seaplane there must be a minimum of 18 inches clearance between the water and the propeller disc.

All aircraft must be designed so that the edge of the propeller disc does not come any closer than one inch to the airframe. This is known as ***radial clearance***.

The propeller must be positioned at least one-half inch in front of, or behind any part of the airframe, other than in the area of the spinner and cowling where only a positive clearance is required.

FAR 23/25.925 gives full details of the propeller clearance requirements.

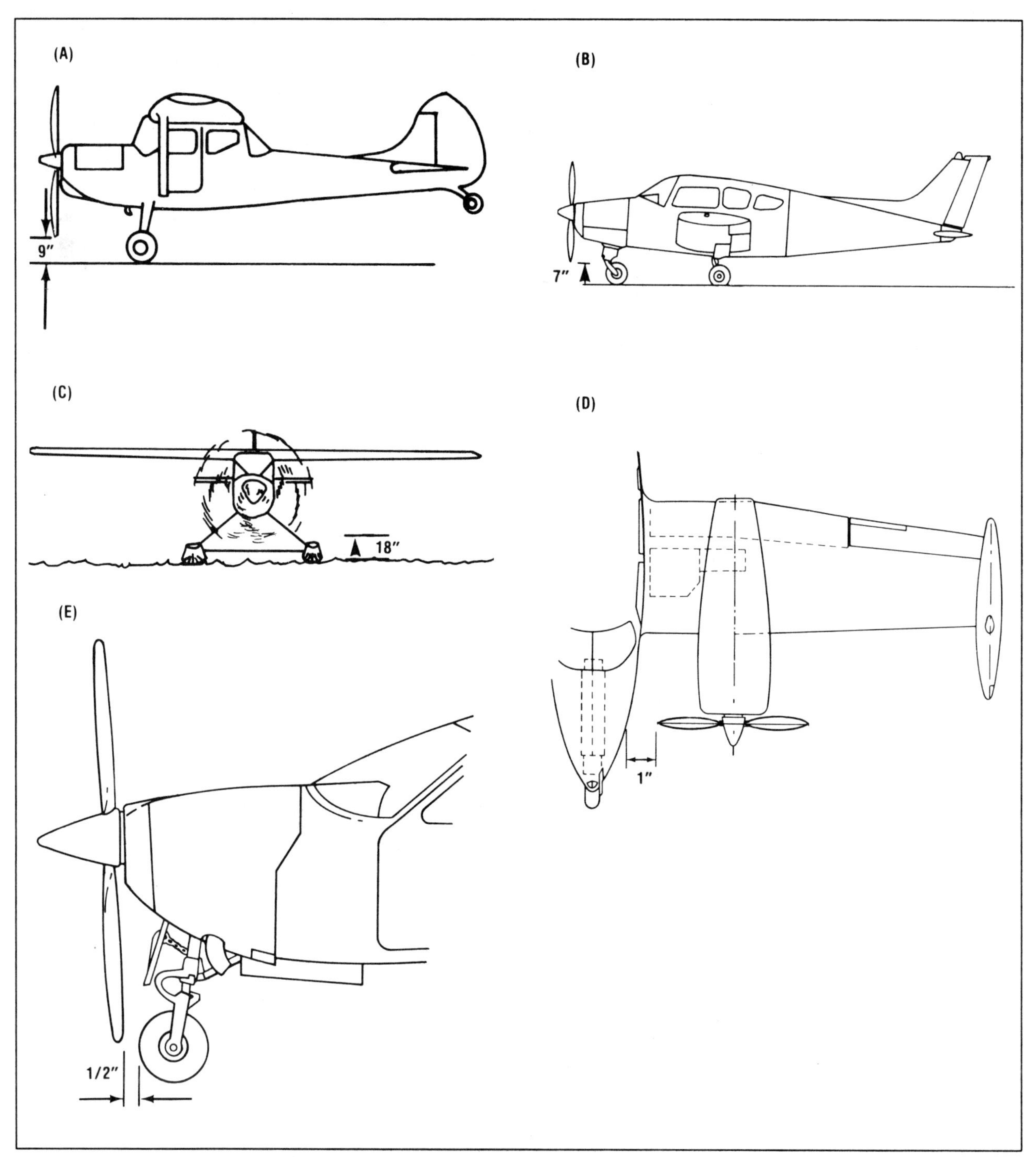

Figure 2-5A. Propeller ground clearance on a tailwheel aircraft.

Figure 2-5B. Minimum ground clearance on a tricycle-geared aircraft.

Figure 2-5C. Minimum propeller water clearance on a seaplane.

Figure 2-5D. Minimum propeller radial clearance.

Figure 2-5E. Minimum longitudinal clearance.

4. Feathering System Requirements

If a propeller can be feathered, there must be some means of unfeathering it in flight (FAR 23/25.1153).

If a propeller system uses oil to *feather* the propeller, a supply of oil must be reserved for feathering use only. A provision must be made in this system

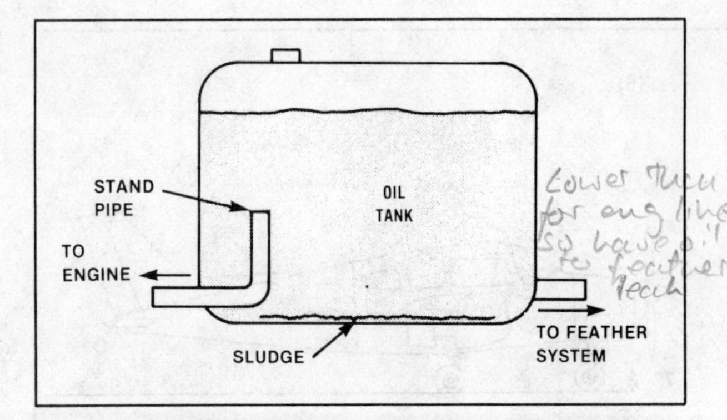

Figure 2-6. Engine oil tank with a standpipe for a system which uses oil to feather the propeller.

to prevent sludge or foreign matter from affecting the feathering oil supply (FAR 23/25.1027). These requirements are normally met by using a standpipe in the engine oil tank with an outlet only to the propeller feathering system.

A separate feathering control is required for each propeller and must be configured to prevent accidental operation (FAR 23/25.1153). This may be done by the use of a separate feathering control such as a feathering button or by requiring an extreme movement of the propeller control.

B. Propeller Maintenance Regulations

1. Authorized Maintenance Personnel

The inspection, adjustment, installation and minor repair of a propeller and its related parts and appliances on the engine are the responsibility of the powerplant mechanic. The powerplant mechanic may also perform the 100-hour inspection of the propeller and related components (FAR 65.87).

A propeller repairman may perform or supervise the major overhaul and repair of propellers and related parts and appliances for which he is certificated. The repair and overhaul must be performed in connection with the operation of a certified repair station, commercial operator, or air carrier (FAR 65.103).

An A&P mechanic who holds an Inspection Authorization may perform the annual inspection of a propeller, but he may not approve major repairs and alterations to propellers and related parts and appliances for return to service. Only an appropriately rated facility, such as a ***propeller repair station***, may return a propeller or accessory to service after a major repair or alteration (FAR 65.81 and 65.91).

2. Preventive Maintenance

The following are types of preventive maintenance that may be associated with propellers and their systems: replacing defective safety wiring or cotter keys; lubrication not requiring disassembly other than removal of nonstructural items such as coverplates, cowlings, and fairings; applying preservative or protective material (paint, wax, etc.) to components when no disassembly is required and the coating is not prohibited or contrary to good practice (FAR Part 43, Appendix A(c)).

3. Major Alterations And Repairs

The following are major propeller alterations when not authorized in the FAA propeller specifications: a change to the blade or hub design; a change in the governor or control design; installation of a governor or feathering system; installation of a propeller ***de-icing system;*** installation of parts not approved for the propeller.

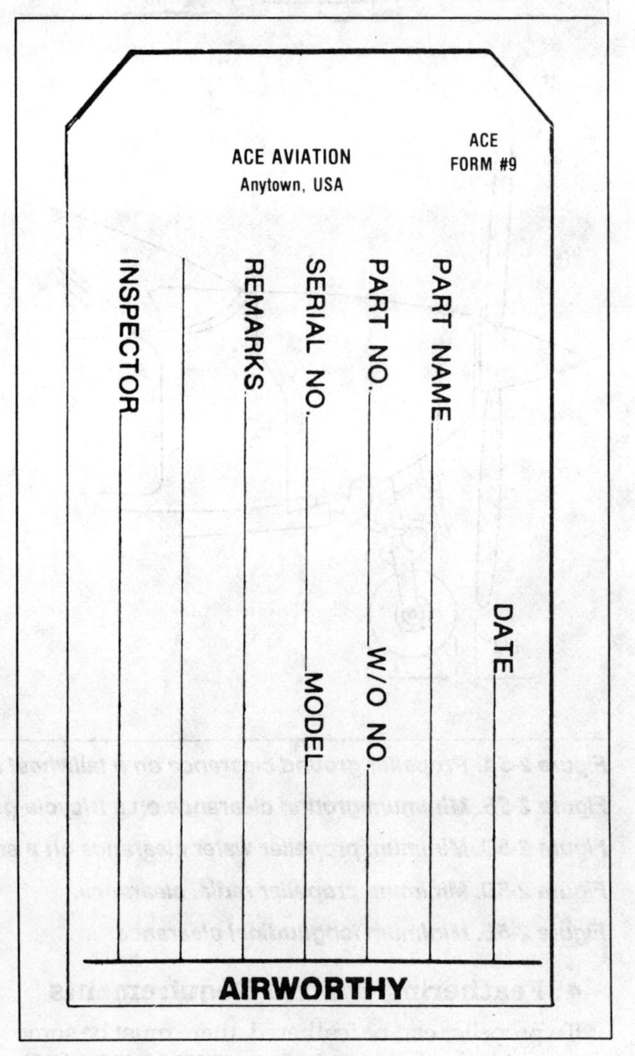
ACE AVIATION
Anytown, USA

ACE
FORM #9

PART NAME ______ DATE ______
PART NO. ______ W/O NO. ______
SERIAL NO. ______ MODEL ______
REMARKS ______
INSPECTOR ______

AIRWORTHY

Figure 2-7. Maintenance release tag.

Propeller major repairs are classified as any repair to, or straightening of steel blades; repairing or machining of steel hubs; shortening of blades; retipping of wood propellers; replacement of outer laminations on fixed-pitch wood propellers; repairing elongated bolt holes in the hub of fixed-pitch wood propeller; inlay work on wood blades; repairs to composition blades; replacement of tip fabric; replacement of plastic covering; repair of propeller governors; overhaul of controllable-pitch propellers; repairs to deep dents, cuts, scars, nicks, etc., and straightening of aluminum blades; the repair or replacement of internal blade elements (FAR Part 43 Appendix A(a)(3) and (b)(3)).

Major repairs and alterations to propellers and control devices are normally performed by the manufacturer or a certified repair station.

When a propeller or control device is overhauled by a repair facility, a maintenance release tag will be attached to the item to certify that the item is approved for return to service. This tag takes the place of a FAA Form 337 and should be attached to the appropriate logbook (FAR Part 43 Appendix B(b)).

4. Annual And 100-Hour Inspections

When performing a 100-hour or annual inspection, Appendix D of FAR 43 specifies that the following areas related to propellers and their controls must be inspected: engine controls for defects, improper travel, and improper safety; lines, hoses, and clamps for leaks, improper condition, and looseness; accessories for apparent defects in security of mounting; all systems for improper installation, poor general condition, defects, and insecure attachment; propeller assembly for cracks, nicks, binds, and oil leakage; bolts for improper torquing and lack of ***safetying;*** anti-icing and de-icing devices for improper operation and obvious defects; control mechanisms for improper operation, insecure mounting, and restricted travel.

These inspections are the minimum required by regulation. Always refer to the manufacturer's manuals for specific inspection procedures.

QUESTIONS:

1. *Which FAR defines the different classes of maintenance for the propeller system?*
2. *What information is required to be on the hub or blade butt of a propeller?*
3. *Why is the static RPM for a fixed-pitch propeller less than tachometer redline?*
4. *Is the propeller control or throttle control in the cockpit used to place the propeller in reverse?*
5. *What is the color of the arc on the tachometer which indicates a critical vibration range?*
6. *What is the minimum ground clearance for the propeller on a tricycle-geared aircraft?*
7. *What are the requirements for a feathering system that uses engine oil to feather the propeller?*
8. *What license or certificate is required to perform or supervise the major repair of a propeller?*
9. *Is the repair of a governor a major or minor repair?*
10. *What maintenance form will be supplied with an overhauled propeller?*

Chapter III
Propeller Theory

As a propeller rotates, it produces lift and causes an aircraft to move forward. The amount of lift produced depends on variables such as engine RPM, propeller airfoil shape, and aircraft speed. The relationship between these variables and the dynamic forces which act upon a rotating propeller will be discussed in this chapter.

A. Propeller Lift And Angle Of Attack

Because a propeller blade is a rotating airfoil, it produces lift by aerodynamic action and pulls an aircraft forward. The amount of lift produced depends on the airfoil shape, RPM, and ***angle of attack*** of the propeller blade sections. Before discussing ways of varying the amount of lift generated by a propeller blade, it is necessary to understand some of the propeller design characteristics.

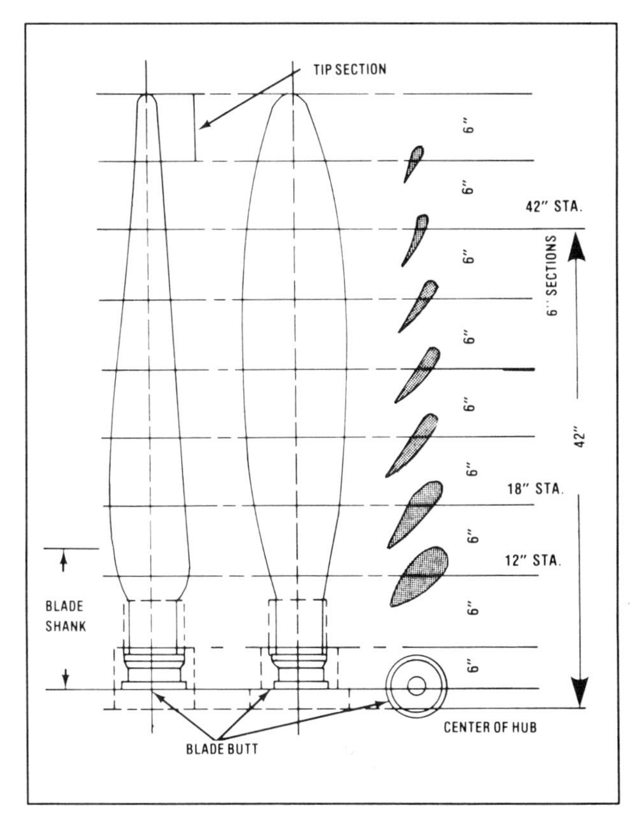

Figure 3-1. Sectioned propeller blade showing pitch distribution, changes in airfoil shape, and blade stations.

Starting from the centerline of the hub of a propeller, each blade can be marked off in one-inch increments known as *blade stations.* If the blade angle is measured at each of these stations, the blade angle near the center of the propeller will be highest with a decrease in blade angle toward the tip. This decrease in blade angle from the hub to the tip is known as *pitch distribution.* A cross section of each blade station will show low-speed airfoils near the hub and high-speed airfoils toward the tip. The pitch distribution and the change in airfoil shape along the length of the blade are necessary because each section is moving at a different velocity with the slowest speeds near the hub and the highest speeds near the tip.

To illustrate the difference in the speed of airfoil sections at a fixed RPM, consider three airfoil sections on a propeller blade. If a propeller is rotating at 1,800 RPM, the ten-inch station will travel 5.25 feet per revolution (107 MPH), while the twenty-inch station must travel 10.5 feet per revolution (214 MPH), and the thirty-inch station has to move 15.75 feet per revolution (321 MPH). The airfoil that gives the best lift at 107 MPH would be inefficient at 321 MPH. Thus the airfoil is changed gradually throughout the length of the blade (Figure 3-2).

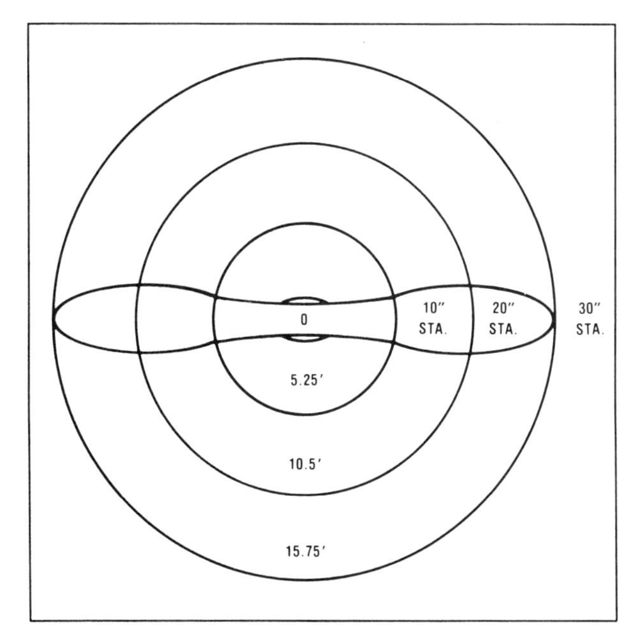

Figure 3-2. Comparative distance of propeller section paths at three blade stations.

A look at one blade section will illustrate how the blade angle of attack on a fixed-pitch propeller can change with different flight conditions. *Angle of attack* is the angle between the airfoil chord line and the relative wind, where the relative wind is a result of the combined velocities of rotational speed (RPM) and airspeed. The following examples will serve to demonstrate how the angle of attack can change:

Example 1: If the aircraft is stationary with no wind and an RPM of 1,200, the propeller blade angle of 20 degrees at the twenty-inch blade station will have an angle of attack of 20 degrees. This is because the relative wind is from the direction opposite to the movement of the propeller.

Example 2: With the same conditions as Example 1, except that the aircraft is moving forward at 50 MPH, the relative wind is now causing an angle of attack of 0.8 degrees.

Example 3: With the same conditions as Example 2, except that the propeller RPM is increased to 1,500 RPM, the relative wind is now causing an angle of attack of 4.4 degrees.

The most desirable angle of attack is between two and four degrees with any angle above 15 degrees being ineffective (airfoils stall at about 15 degree angle of attack). Fixed-pitch propellers may be selected to give this two to four degree angle of attack at either climb or cruise airspeeds and RPM, depending on the desired flight characteristics of the aircraft.

B. Forces Acting On The Propeller

As a propeller rotates, many forces are interacting causing tension, torsion, compression, and bending stresses which the propeller must be designed to withstand. These stresses are broken down into six forces.

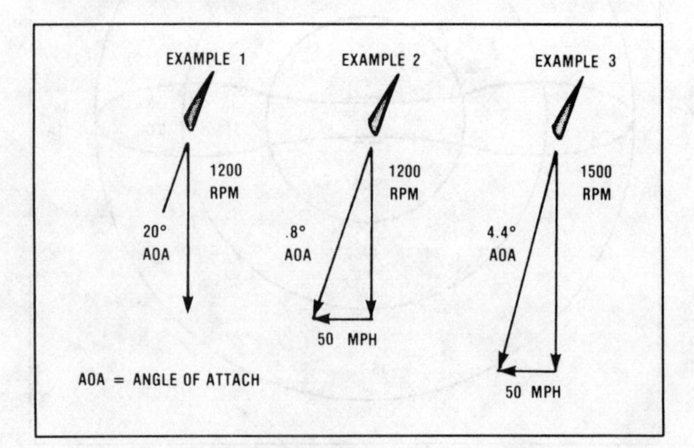

Figure 3-3. Variations in blade angle of attack with different airspeeds and RPM.

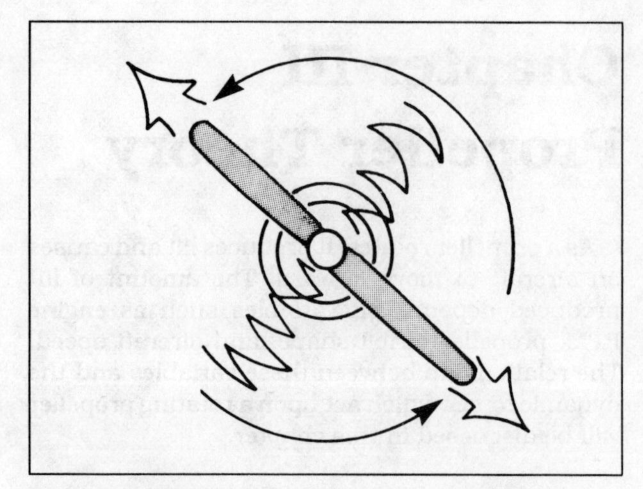

Figure 3-4. Centrifugal force tries to pull the blades out of the hub.

1. The Five Operational Forces

a. Centrifugal Force

The force which causes the greatest stress on a propeller is ***centrifugal force.*** Centrifugal force can best be described as the force which tries to pull the blades out of the hub. The amount of stress created by centrifugal force may be greater than 7,500 times the weight of the propeller blade.

b. Thrust Bending Force

Thrust bending force tends to bend the propeller blades forward at the tips because the lift toward the tip of the blade flexes the thin blade sections forward. Thrust bending force opposes centrifugal force to some degree. By *tilting* the blades forward to the operational position during manufacture, this opposition in forces can be used in some propeller designs to reduce the operational stress.

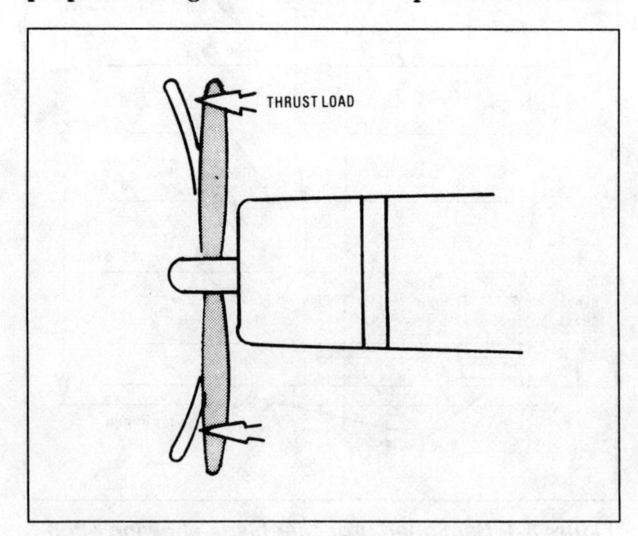

Figure 3-5. Thrust bending forces tend to bend the blades forward.

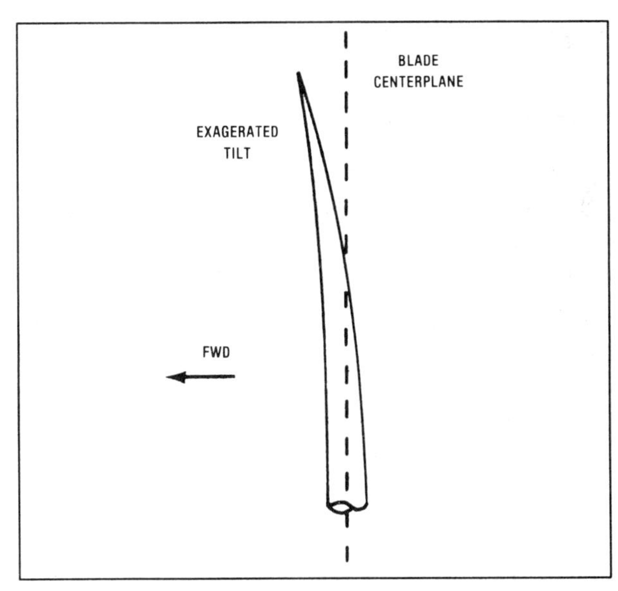

Figure 3-6. A tilted blade is designed to reduce the stress caused by centrifugal and thrust bending forces.

c. Torque Bending Force

Torque bending force is a force which tends to bend the propeller blade back in the direction opposite to the direction of rotation.

d. Aerodynamic Twisting Moment

Aerodynamic twisting moment tries to twist a blade to a higher angle by aerodynamic action.

This results from the center of rotation of the blade being at the mid-point of the chord line while the center of lift (also called center of pressure) is more toward the leading edge of the blade. This tends to cause an increase in blade angle. ***Aerodynamic twisting moment*** is more apparent at higher blade angles of attack and is used in some designs to aid in feathering the propeller.

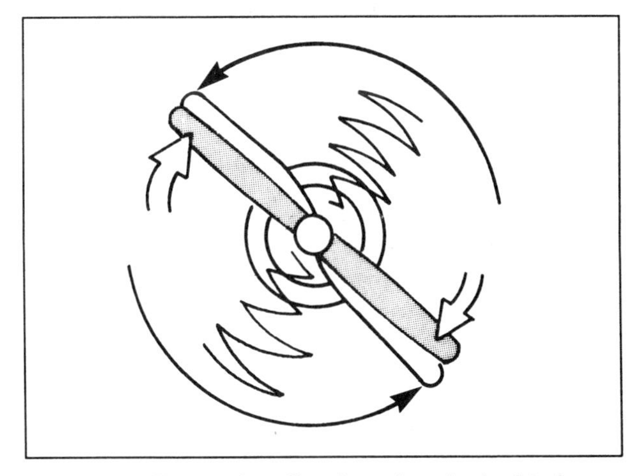

Figure 3-7. Torque bending force bends the blade back against the direction of rotation.

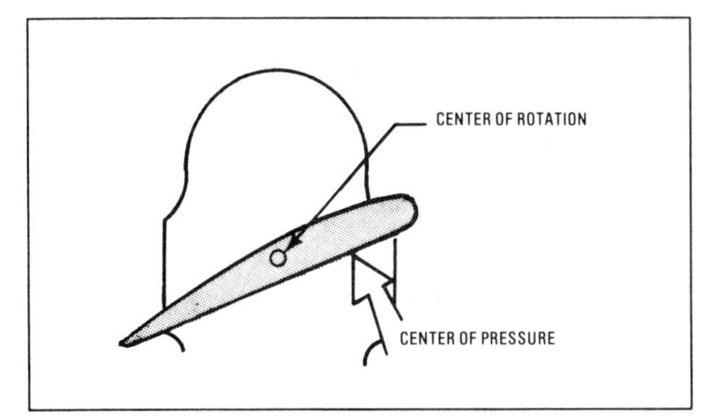

Figure 3-8. Aerodynamic twisting moment tends to increase blade angle.

e. Centrifugal Twisting Moment

Centrifugal twisting moment tends to decrease blade angle and opposes aerodynamic twisting moment. The tendency to decrease blade angle is caused by all parts of a rotating propeller trying to move in the same plane of rotation as the blade centerline. This force is greater than the aerodynamic twisting moment at operational RPM and is used in some designs to cause a decrease in blade angle.

2. Vibrational Force And Critical Range

When a propeller is producing thrust, aerodynamic and mechanical forces are present which cause the blades to vibrate. If not compensated for in the design, these vibrations may cause excessive flexing, work-hardening of the metal, and result in sections of the propeller blade breaking off during operation.

Aerodynamic forces have a great vibration effect at the tip of a blade where the effects of transonic speeds cause buffeting and vibration. These vibrations may be decreased by use of the proper airfoils and tip designs.

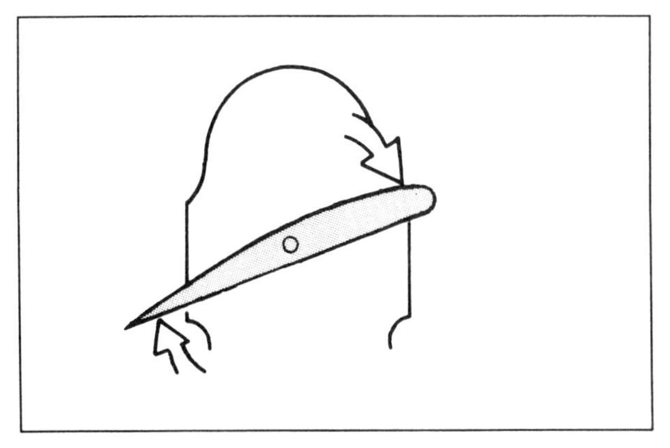

Figure 3-9. Centrifugal twisting moment tends to decrease blade angle.

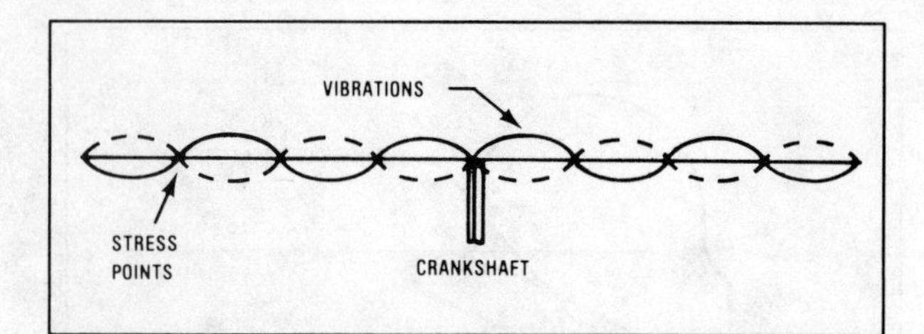

Figure 3-10. Power pulses from the engine cause the propeller to vibrate.

Mechanical vibrations are generated by the power pulses in a piston engine and are considered to be more destructive in their effect than aerodynamic vibration. These engine power pulses cause a propeller blade to vibrate and set up standing wave patterns that cause metal fatigue and failure. The location and number of stress points changes with different RPM settings, but the most critical location for these stress concentrations is about six inches in from the tip of the blades.

Most airframe-engine-propeller combinations have no problem in eliminating the detrimental effects of these vibrational stresses. However, some combinations are sensitive to certain RPM ranges and have this ***critical range*** indicated on the tachometer by a red arc. The engine should not be operated in the critical range except as necessary to pass through it to set a higher or lower RPM. If the engine is operated in the critical range, there is a possibility of structural failure in the aircraft due to the vibrational stresses set up.

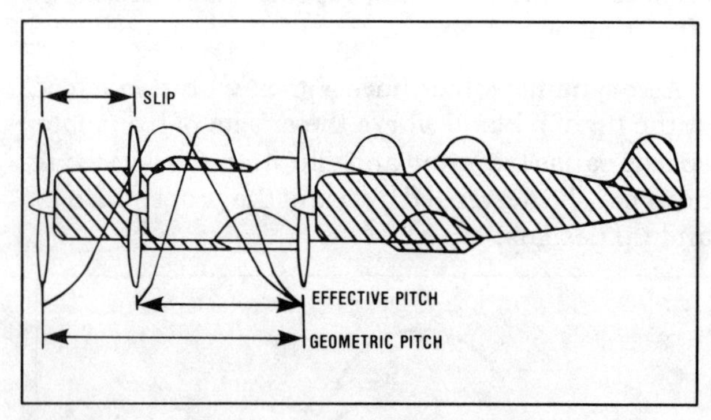

Figure 3-11. Effective and geometric pitch.

C. Propeller Pitch

Propeller pitch is defined as the distance in inches that a propeller will move forward in one revolution. This is based on the propeller blade angle at the 75% blade station. As defined, propeller pitch is more properly called ***geometric pitch*** and is theoretical in that it does not take into account any losses due to inefficiency.

Effective pitch is the distance that an aircraft actually moves forward in one revolution of the propeller. Effective pitch may vary from zero, when the aircraft is stationary on the ground, to about 85% during the most efficient flight conditions.

The difference between geometric pitch and effective pitch is called ***slip***.

As an example: if a propeller is said to have a pitch of 50 inches, in theory it will move forward 50 inches in one revolution. But, if the aircraft actually only moves forward 35 inches in one revolution, then the effective pitch is 35 inches and its pitch efficiency is 70%. Slip then, is 15 inches or a 30% loss of efficiency.

QUESTIONS:

1. *Is the lowest propeller blade angle near the shank or the tip?*
2. *What is pitch distribution?*
3. *With a fixed RPM, does propeller blade angle of attack increase or decrease as airspeed increases?*
4. *What is the most desirable blade angle of attack?* 2-4
5. *What is the greatest force acting on a propeller?*
6. *Why are some propeller blades tilted?*
7. *What force tends to decrease propeller blade angle?*
8. *Where is the most critical location for vibrational stresses on a propeller blade?*
9. *How is critical range indicated in the cockpit?*
10. *What is effective pitch?*

Chapter IV
Fixed-Pitch Propellers And Propeller Blades

This chapter deals with the construction and maintenance of fixed-pitch propellers and the blades of changeable-pitch propellers. The information presented here is general in nature and gives representative values for repair dimensions.

Always refer to the manufacturer's manual for specific information about a particular propeller.

A. Wood Propellers And Blades

Wood propellers are often found on older single-engine aircraft using fixed-pitch propellers and on some controllable-pitch installations on vintage aircraft. Most have a natural wood finish, with some designs using a black or gray plastic coating.

1. Construction

Wood propellers are made of several layers of wood bonded together with a waterproof resin glue. The woods most commonly used are mahogany, cherry, black walnut, oak, and birch (birch being the most widely used). Each layer of a propeller is usually of the same thickness and type of wood. A minimum of five layers of wood are used. The planks of wood glued together form a *blank*.

During fabrication the blank is roughed to shape and is allowed to set for a week to allow equal distribution of moisture through all of the layers. The rough-shaped blank is called a *white*.

The white is finished to the exact airfoil and pitch dimensions for the desired performance characteristics. During this process the center bore and bolt holes are drilled.

From this point, the *tip fabric* is applied on the propeller. The cotton tip fabric is glued to the last 12 to 15 inches of the propeller blade and serves

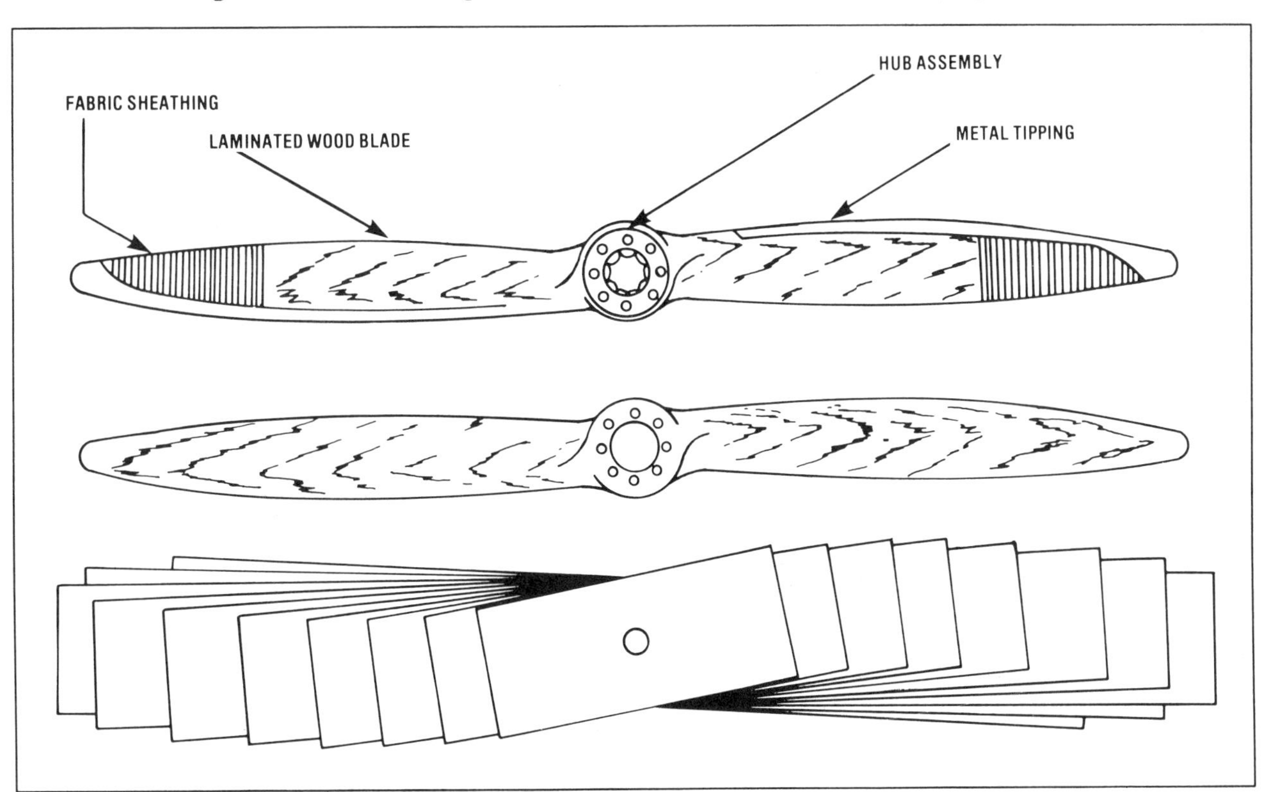

Figure 4-1. Three stages of wood propeller production: glued planks, white, and finished propeller.

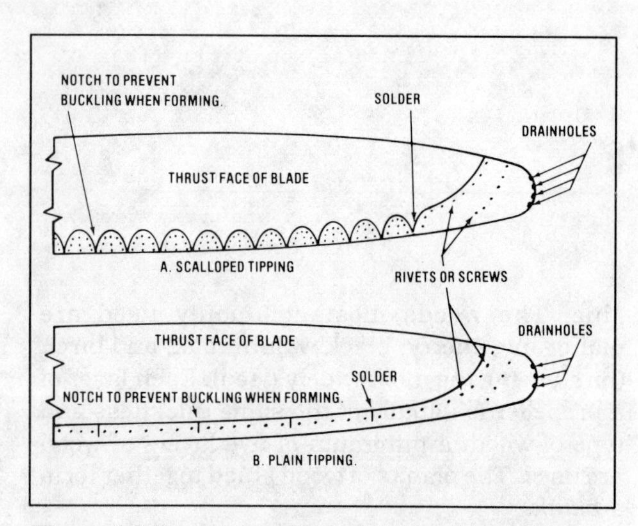

Figure 4-2. Two styles of metal tipping installations.

to reinforce the strength of the thin sections of the tip. The fabric is doped to prevent deterioration by weather and the sun's rays.

The propeller is varnished with a coat of water-repellant clear varnish to protect the wood surface.

Metal tipping is applied to the leading edge of the propeller to prevent damage from small stones during ground operations. The tipping is made of monel, brass, terneplate, or stainless steel. The metal is shaped to the leading edge contour and is attached to the blade by countersunk screws in the thick blade sections and copper rivets in the thin sections near the tip. The screws and rivets are safetied into place with solder.

Three #60-size holes, 3/16-inch deep, are drilled in the tip of each blade to release moisture from the propeller and allow the wood to breathe. The propeller is then balanced and the finish coats of varnish are applied.

Some propellers do not use tip fabric, but are coated with plastic before the metal tipping is applied. This plastic coating provides protection and added strength to the propeller.

Wood blades for controllable-pitch propellers are constructed in the same manner as fixed-pitch propellers except that the blade is placed in a metal sleeve at the shank and secured with lag screws.

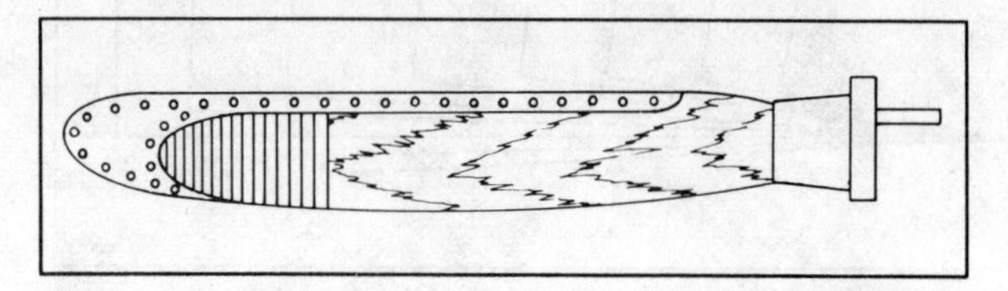

Figure 4-3. Wood blade for a constant-speed propeller.

2. Inspection, Maintenance And Repair

Wood propellers are made of many components and require a close inspection of each part to assure proper operation and prevent failures.

Defects that may occur in the wood include separation of laminations, dents or bruises on the surface (especially the face), scars across the blade surface, broken sections, warping, worn or oversize centerbore and bolt holes.

Separation of laminations is not repairable unless it occurs at the outside lamination of fixed-pitch propellers. Delamination of the outer layers may be repaired by a repair station.

Dents, bruises, and scars on the blade surfaces should be inspected with a magnifying glass while flexing the blade to expose any cracks. Cracks that show could cause failure and may be repaired by an inlay at a repair facility. Defects that have rough surfaces or shapes that will hold a filler, but will not induce failure, may be filled with a mixture of glue and clean fine sawdust. To apply this mixture, thoroughly work and pack the mixture into the defect, allow it to dry, and then sand the surface smooth and refinish with varnish. It is very important that all loose or foreign material be removed from the damaged area to insure that the glue will adhere to the wood.

Small cracks that are parallel to the grain may be repaired by working resin glue into the crack. When the glue is dry, the area is sanded smooth and refinished with varnish. Small cuts are treated in the same manner.

Broken sections may be repaired by a repair facility, depending on the location and severity of the break.

Worn or oversize bolt holes may be repaired by the use of inserts to restore the original dimensions. This repair is performed by a repair station and is subject to wear limits.

The tip fabric covering should be checked for cracks or bubbles in the material, chipping of the paint, and wrinkles that appear when the tip is twisted or flexed.

If the tip fabric has surface defects of three-quarter inch or less in diameter (and not an indication of a breakdown in the wood structure), the defect may be filled with lacquer. Several coats of lacquer are applied until the defect blends in with the fabric surface. Defects larger than 3/4 of an inch should be referred to a repair facility.

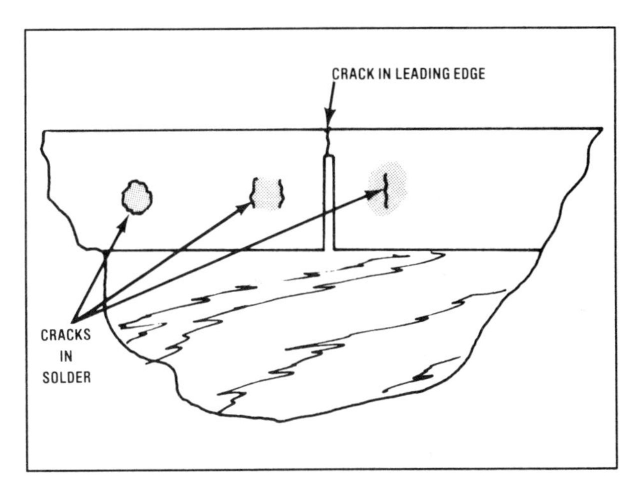

Figure 4-4. Inspect solder safeties for cracks.

When inspecting the metal tipping, look for looseness or slipping, loose screws or rivets, cracks in the solder joints, damage to the metal surface, and cracks in the metal, especially on the leading edge.

If the tipping is loose or slipping, refer the propeller to an ***overhaul facility*** for repairs as this is an indication of wood deterioration. Loose screws and rivets are indicated by a small crack appearing in the solder over the screw or rivet and normally running parallel to the chord line of the blade. In advanced stages the crack will form a circle around the screw head or rivet head. The solder which safeties the screw or rivet should be removed by using a heavy duty soldering iron to melt the solder. Then brush the solder away with a steel brush. Care must be taken to heat the solder quickly and not allow the wood to be scorched in the process. Tightened a loose screw or replace it with the next larger size screw as necessary. Repeen or replace a loose rivet. Resafety the screw or rivet with solder. The surface is filed, smooth sanded, and varnished.

Cracks in the solder joint near the blade tip may be indications of wood deterioration. Inspect the area closely while flexing the blade tip. If no defects are found, the joint may be resoldered, but inspect the area closely at each opportunity for evidence of recurrence.

Damage to the metal surface normally is in the form of dents and scratches from stone strikes. Inspect the damage and the wood in the damaged area for evidence of defects. If none are found, the dent may be filled with solder, filed smooth, and varnished. If a crack in the metal is found, inspect the area carefully for any further damage and, if none is found, stop-drill the crack, being careful not to enter the wood more than necessary. Fill the crack with solder.

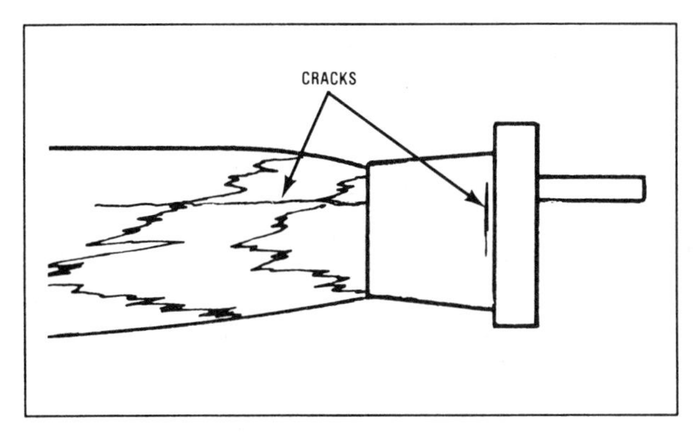

Figure 4-5. Inspect the metal sleeve area of a wood constant-speed blade for cracks.

On a changeable-pitch blade, check the metal sleeve and the wood next to the sleeve for cracks. This may indicate loose or broken lag screws and should be referred to an overhauled facility for correction.

If the varnish should begin to peel or chip, the surface can be sanded lightly to feather-in the edges of the irregularity and then apply a fresh coat of varnish to the area.

In addition to the repairs mentioned above, a repair station can perform the following major repairs: replacement of metal tipping; replacement of tip fabric; and replacement of plastic coatings.

The following defects are *not* repairable and are reasons for considering a propeller unairworthy: a crack or deep cut across the grain; a split blade; separated laminations, except for the outside laminations of a fixed-pitch propeller; unused screw or rivet holes; any appreciable warp; an appreciable portion of wood missing; cracks, cuts, or damage to the metal sleeve of a changeable-pitch propeller; an oversized crankshaft bore in a fixed-pitch propeller; cracks between crankshaft holes and bolt holes; cracked internal laminations; excessively elongated bolt holes; and broken lag screws in a changeable-pitch propeller.

If balancing equipment is available, the propeller balance may be checked and corrected as follows: place the propeller on the balancing stand and check for balance in the horizontal and vertical positions. This should be done in an area free from drafts and wind currents that may cause the propeller to appear out of balance.

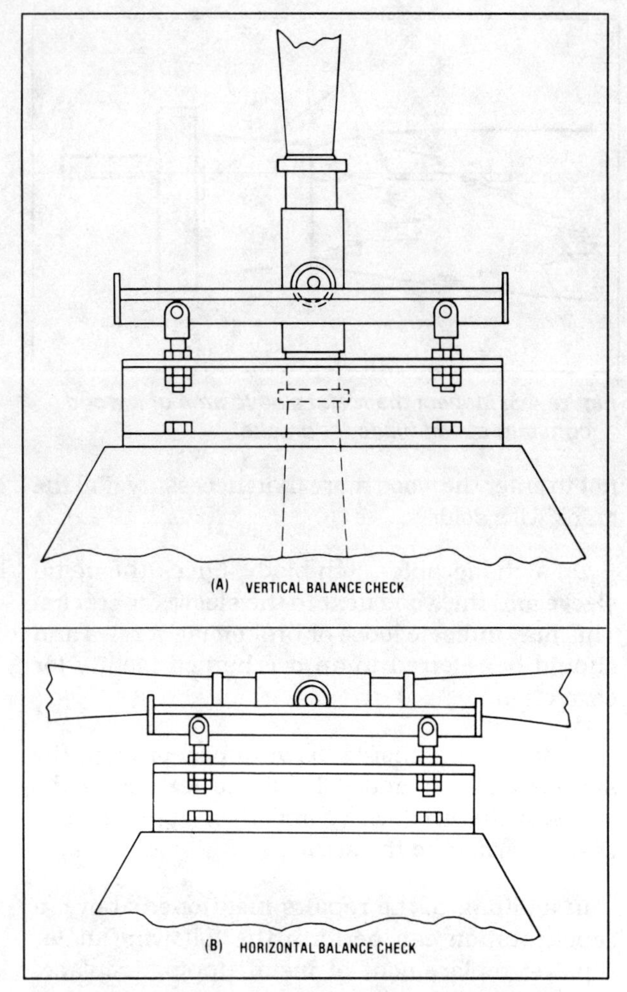

Figure 4-6. Positions of a two-bladed propeller during a balance check.

If the propeller is out of balance in the horizontal position, while the propeller is on the balancing stand, bend short lengths of solder (two to three inches) and hang them over the tip of the light blade until the propeller is balanced horizontally. Remove the solder pieces from the tip and set them aside. Place the propeller in a working fixture with the face up. Remove the varnish from the metal tip cap and use a large soldering iron to melt the solder pieces onto the tip cap. Smooth out the solder with the

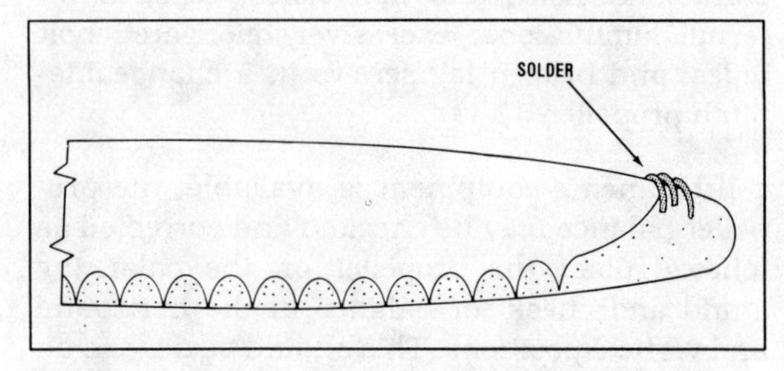

Figure 4-7. Adjusting horizontal balance by the use of solder pieces.

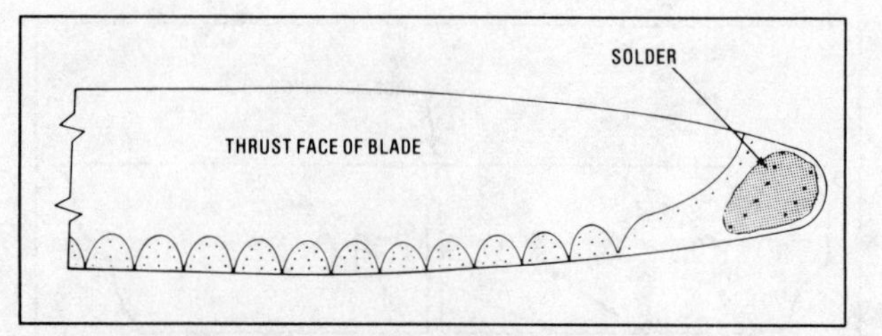

Figure 4-8. Correct horizontal balance by spreading solder on the face of the tip.

iron to a thin even coating. Allow the solder to cool and then file and sand the solder to a smooth, even finish. Varnish the worked area and then recheck the horizontal balance.

A slight horizontal imbalance can be corrected by applying a coat of varnish to the light blade, but the varnish must be allowed to dry for 48 hours before rechecking the balance.

If a vertical imbalance is noted, leave the propeller on the balance stand and stick clay or putty to the light side of the ***boss*** to determine the amount of weight needed to bring the propeller into balance. Remove the putty, weigh it, and prepare a lead or brass plate of slightly more weight than the putty. The plate is attached to the boss with four screws with the edges of the plate beveled as necessary to remove excess weight and bring the propeller into vertical balance. The plate is then varnished.

If a wood propeller is placed in storage, it should be placed in a horizontal position to maintain an even moisture distribution throughout the wood. The storage area should be cool, dark, dry, and well ventilated. Do not wrap the propeller in any material that will seal off the propeller from the surrounding air flow or the wood will rot.

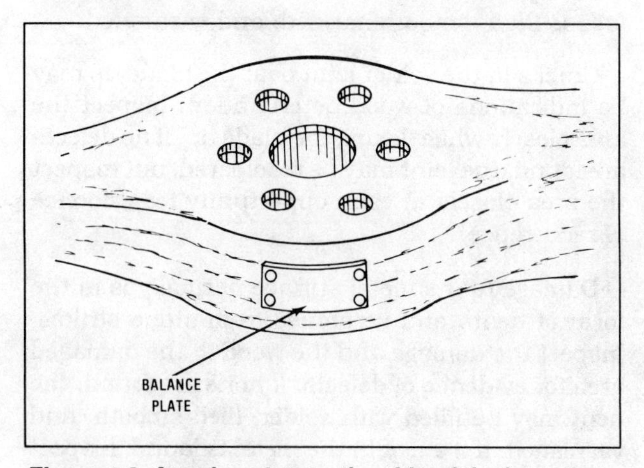

Figure 4-9. Attach a plate to the side of the boss to correct vertical balance.

B. Aluminum Propellers And Blades

Aluminum propellers are the most widely used type of propellers in aviation. Aluminum propellers are more desirable than wood propellers because thinner, more efficient airfoils may be used without sacrificing structural strength. Better engine cooling is also achieved by carrying the airfoil sections close to the hub and directing more air over the engine. These propellers require much less maintenance than wood propellers, thereby reducing the operating cost.

1. Construction

Aluminum propellers are made of aluminum alloys and are finished to the desired airfoil shape by machine and manual grinding. The pitch is set by twisting the blades to the desired angles.

As the propeller is being finished by grinding, its balance is checked and adjusted by removing metal from the tip of the blade to adjust horizontal balance and the boss or leading and trailing edges of the blades to adjust vertical balance. Some fixed-pitch propeller designs have their horizontal balance adjusted by placing lead wool in balance holes near the boss and their vertical balance corrected by attaching balance weights to the side of the boss.

Once the propeller is ground to the desired contours and the balance is adjusted, the surfaces are finished by plating, chemical etching, and/or painting. Anodizing is the most commonly used finishing process.

2. Inspection, Maintenance And Repair

As mentioned previously, an advantage of aluminum propellers is the low cost of maintenance. This is due to the one-piece construction and the hardness of the metal from which the propellers are made. However, any damage that does occur is critical and may result in blade separation. For this reason, the blades must be inspected carefully and any damage must be repaired as soon as possible.

Before a propeller is inspected it should be cleaned with a solution of mild soap and water to remove dirt, grass stains, etc.

The propeller blades should be inspected for pitting, nicks, dents, cracks, and corrosion, especially on the leading edge and face. A four-power magnifying glass will aid in these inspections. A dye penetrant inspection should be performed if cracks are suspected. The condition of the paint should also be noted.

A majority of the surface defects that occur on the blades can be repaired by the powerplant mechanic. Defects on the leading and trailing edge of a blade may be dressed out by the use of needle files. The repair should blend in smoothly with the edge and should not leave any sharp edges or angles. The approximate maximum allowable size of a repaired edge defect is 1/8 inch deep and no more than 1 1/2 inches long. Repairs to the face and back of a blade are performed with a spoon-like riffle file which is used to dish out the damaged area. Permissible reductions in blade thickness and width as noted in the manufacturer's publications or AC 43.13-1A must be observed. The repairs are finished by polishing with very fine sand paper, moving the paper in a direction along the length of the blade, and then

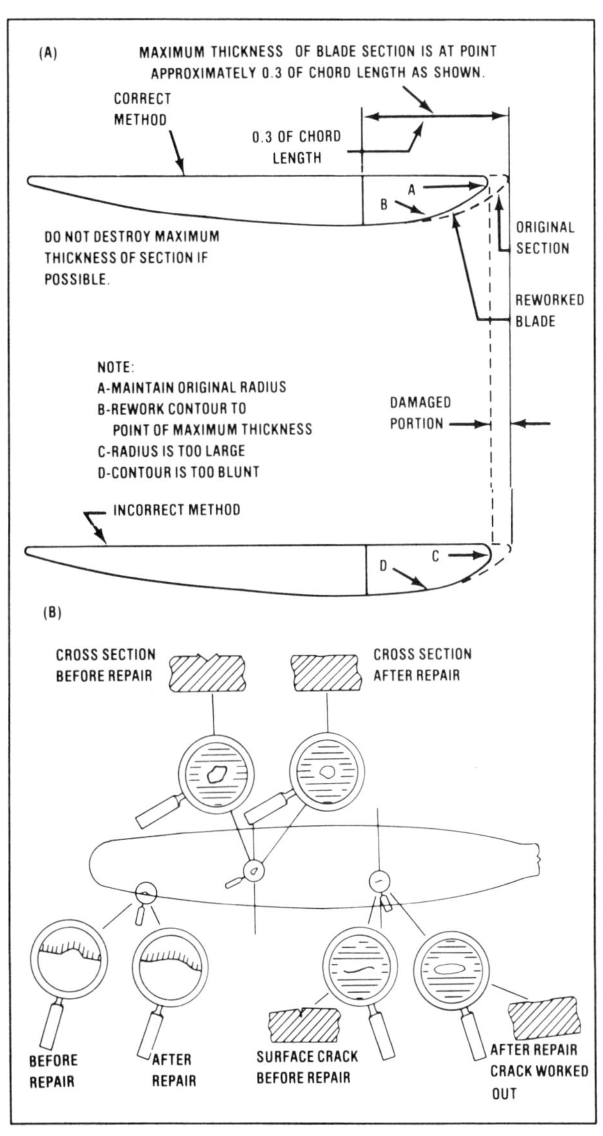

Figure 4-10A. Method of repairing leading edge damage.

Figure 4-10B. Before and after illustrations of defects.

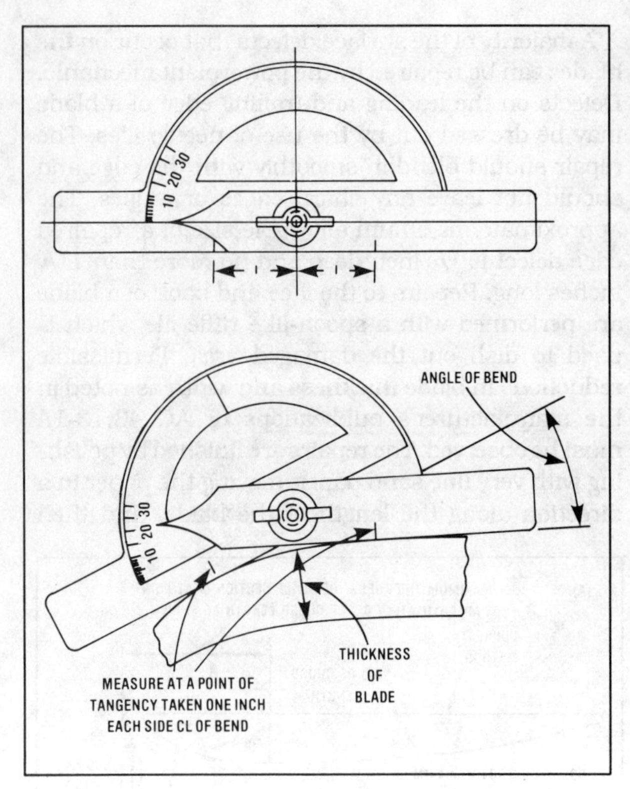

Figure 4-11. Measuring blade bend angle with a protractor.

treating the surface with Alodine®, paint, or other appropriate protective coating. The repair dimensions presented here are for example only. The allowable repair dimensions may be different depending on the model propeller being repaired.

The boss should be inspected for damage and corrosion inside the center bore and on the surfaces which mount on the crankshaft. The bolt holes should be inspected for damage, corrosion, and proper dimension. Dowel pins should be inspected for damage, security, and dimension.

Light corrosion in the boss can be cleaned with sandpaper and then painted or treated to prevent the recurrence of corrosion. Propellers with damage or heavy corrosion in the boss area should be referred to a repair station for appropriate repairs. Dimensional wear in the boss area should also be referred to a repair facility.

Damage in the shank area of a propeller blade should be referred to an overhaul facility for corrective action. All forces acting on the propeller are concentrated on the shank and any damage in this area is critical.

If a blade has been bent, the angle of the bend and the blade station of the bend center can be measured and, by using the proper chart, a determination can

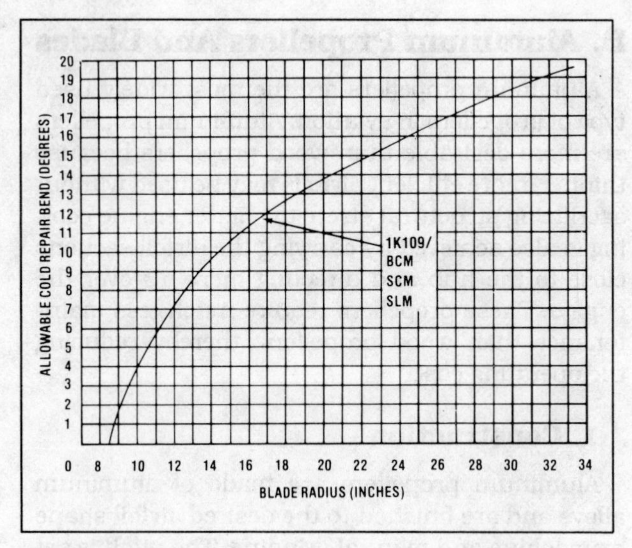

Figure 4-12. Chart for maximum bend allowed in a propeller blade.

be made as to the repairability of the blade. To make this decision, determine the center of the bend and measure from the center of the boss to determine the blade station of the center of the bend. Next, mark the blade one inch on each side of the bend centerline and measure the degree of bend by using a protractor similar to the one shown in Figure 4-11. (Be sure to have the protractor tangent to the one-inch lines when measuring the angle.) Use the appropriate chart to determine if the bend is repairable. When reading the chart, anything above the graph line is not repairable. If the proper chart is not available, take the measurements and contact an overhaul facility for a decision before sending the propeller to the facility for straightening.

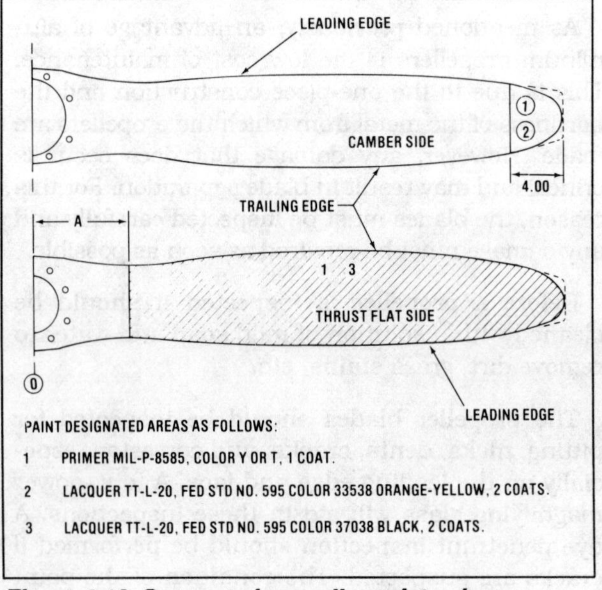

Figure 4-13. Suggested propeller paint scheme.

After the propeller has been repaired, the surfaces may have to be repainted. The face of each blade should be painted with one coat of zinc chromate primer and two coats of flat black lacquer from the six-inch station to the tip. The back of the blade should have the last four inches of the tip painted with one coat of zinc chromate and two coats of a high visibility color such as red, yellow, or orange. The color scheme on the back of the blade on some aircraft differs from that described here, so the original color scheme may be duplicated if desired.

Repairs and modifications that may be performed by a repair station, not mentioned above, includes the removal of deep and large surface defects, shortening of blades, and changing the pitch of blades.

C. Steel Propellers And Blades

Steel propellers and blades are found primarily on antiques and transport aircraft. These are normally of hollow construction. The primary advantage of the hollow blades is in the reduced weight. Steel blades, whether solid or hollow are very durable and resistant to damage.

1. Construction

Solid steel propellers are forged and machined to the desired contours and the proper pitch is achieved by twisting the blades.

One method of constructing hollow steel blades is by assembling a rib structure, attaching steel sheets to the structure, and filling the outer sections of the blade with a foam material to absorb vibration and maintain a rigid structure.

2. Inspection, Maintenance And Repair

Steel blades are not as susceptible to damage as aluminum or wood blades, but any damage is critical due to the brittle metals used. Consequently, damage must be located and corrected as soon as possible.

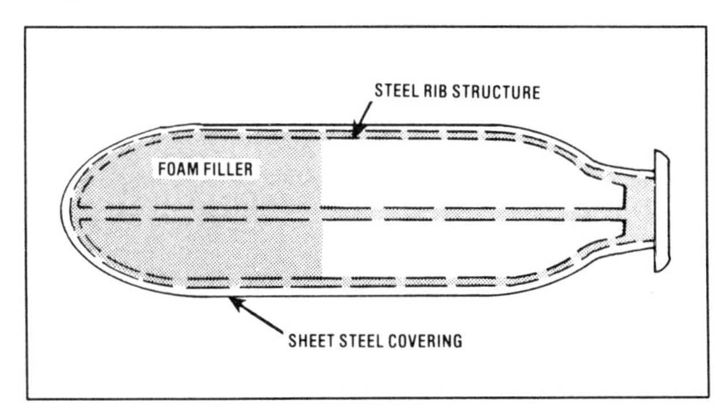

Figure 4-14. Hollow steel blade construction.

A visual inspection may be performed with the aid of a magnifying glass and the use of a dye penetrant. Magnetic particle inspections may be performed according to the manufacturers specifications.

Bent blades should be treated like aluminum blades with the location and amount of bend being used to determine if the blade is repairable.

All repairs to steel propellers and blades, including slight dents and nicks, are major repairs and must be performed by a repair station.

D. Fixed-Pitch Propeller Designation Systems

Two propeller designation systems are covered so that the mechanic will be able to understand the systems and notice the differences in propeller designs by their designation. The McCauley and Sensenich systems covered are representative of those presently in use.

1. McCauley Designation System

A McCauley propeller designation 1B90/CM7246 has a basic design designation of 1B90. The CM component of the designation indicates the type of crankshaft the propeller will fit, blade tip contour, adapter used, and provides other information pertaining to a specific aircraft installation. The 72 indicates the diameter of the propeller in inches and the 46 indicates the pitch of the propeller at the 75% station.

2. Sensenich Designation System

The Sensenich designation 76DM6S5-2-54 indicates a propeller with a designed diameter of 76 inches. The *D* designates the blade design and the *M6* indicates hub design and mounting information (bolt hole size, dowel pin location, etc.). The *S5* designates the thickness of the spacer to be used when the propeller is installed. The *2* indicates that the diameter has been reduced two inches from the designed diameter meaning that this propeller has an actual diameter of 74 inches. The *54* designates the pitch, in inches, at the 75% station.

In either designation system, a change in pitch will be indicated by the pitch stamping on the hub being restamped to indicate the new pitch setting.

Other propeller manufacturers use designation systems that are similar to the McCauley and Sensenich systems.

QUESTIONS:

1. *What type of wood is most commonly used in constructing wood propeller blades?*
2. *What is the purpose of the tip fabric on a wood propeller blade?*
3. *What is the purpose of the holes in the tip of a wood propeller?*
4. *How can loose screws and rivets be detected on metal tipping?*
5. *If the outer lamination of a wood fixed-pitch propeller is starting to separate, can it be repaired or must the propeller be scrapped?*
6. *What is the proper way to store a wood propeller?*
7. *List some advantages of aluminum propellers over wood propellers.*
8. *Why must surface defects be repaired promptly on a metal propeller?*
9. *What solution should be used to clean a metal propeller?*
10. *On which part of a propeller blade are all the stresses concentrated?*
11. *What type of repair is the removal of a slight scratch on a steel blade?*
12. *Break down the following designation: 1C172/DM7553.*

Chapter V
Propeller Installations

The three types of propeller installations — ***flanged shaft, tapered shaft,*** and ***splined shaft*** — are discussed in this chapter. Although the chapter discusses fixed-pitch propeller installations, the principles are the same as for other types of propeller installations (constant-speed, reversing, etc.) and only major variations will be covered in future chapters.

A. Flanged-Shaft Installations

Flanged propeller shafts are found on horizontally opposed and some turboprop engines. The front of the crankshaft is formed into a flange four to eight inches across and 90 degrees to the crankshaft centerline. Mounting bolt holes and dowel pin holes are machined into the flange and, on some flanges, threaded inserts are pressed into the bolt holes.

1. Preparation For Installation

Before the propeller is installed, the flange should be inspected for corrosion, nicks, burrs, and other surface defects. The defects should be repaired in accordance with the engine manufacturer's recommendations. Light corrosion can be removed with very fine sandpaper. If a bent flange is suspected, a *run-out* inspection should be performed on the crankshaft flange. The bolt holes and threads of inserts should be clean and in good condition.

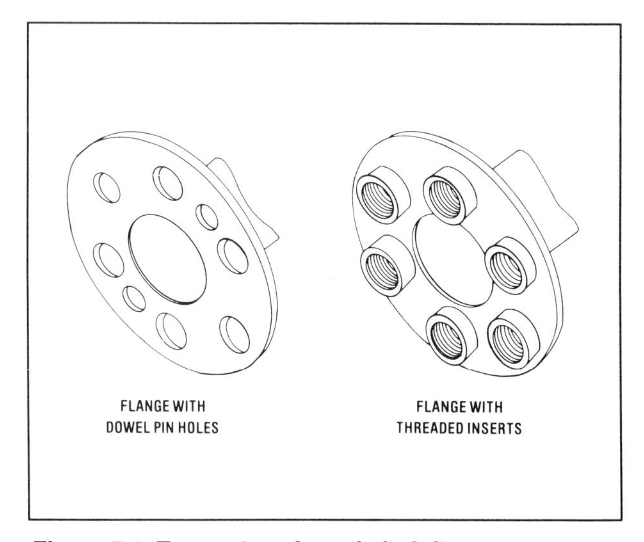

Figure 5-1. Two styles of crankshaft flanges.

With the flange area clean and smooth, a light coat of oil or anti-seize compound is applied to prevent corrosion and allow easy removal of the propeller.

The mounting surfaces of the propeller should be inspected and prepared in a manner similar to that used with the flange.

REINSTALLING - USE NEW ONES

The bolts to be used should be in good condition and inspected for cracks with a dye-penetrant or magnaflux process. Washers and nuts should also be inspected and new fiber locknuts should be used if required in the installation.

2. Installation

The propeller is now ready to mount on the crankshaft. If dowel pins are used, the propeller will fit on the shaft in only one position. If no dowel pins are used, the propeller should be installed in the position called for in the aircraft or engine maintenance manual. Propeller installation position is *critical* for maximum engine life in some installations. If no position is specified on a four-cylinder horizontally-opposed engine, the propeller should be installed so the blades

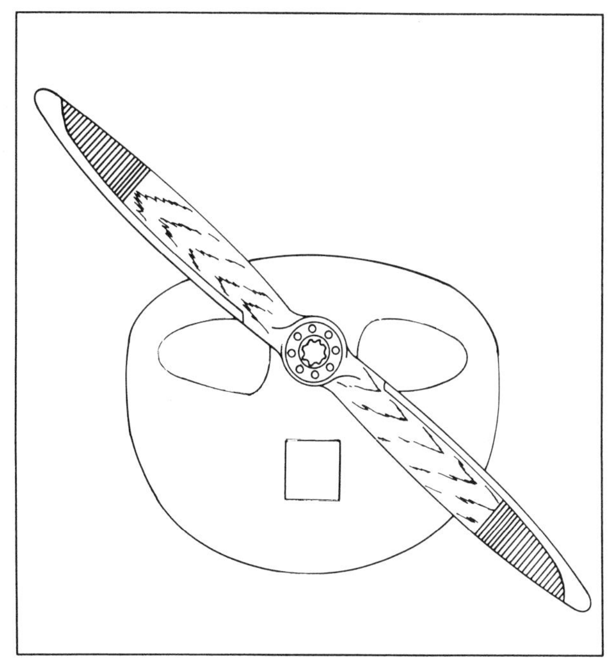

Figure 5-2. Propeller installation position for a four-cylinder engine.

are at the ten o'clock and four o'clock position when the engine stops. This reduces vibration in many instances and puts the propeller in position for hand-propping the airplane.

The bolts, washers, and nuts are installed next, according to the particular installation. The bolts should be tightened slightly. Use an alternating torquing sequence to tighten the bolts to the desired value. Refer to the appropriate manufacturers service information for specific values of propeller torque.

When a *skull cap* spinner is used, the mounting bracket is installed with two of the propeller mounting bolts. If a full spinner is used, a rear bulkhead is installed on the flange before the propeller is installed, and a front bulkhead is installed on the front of the boss before the bolts are placed through the propeller boss. The spinner is now installed using screws.

If a wood propeller is being installed, a faceplate is normally placed on the front of the propeller boss before installing the bolts. The faceplate

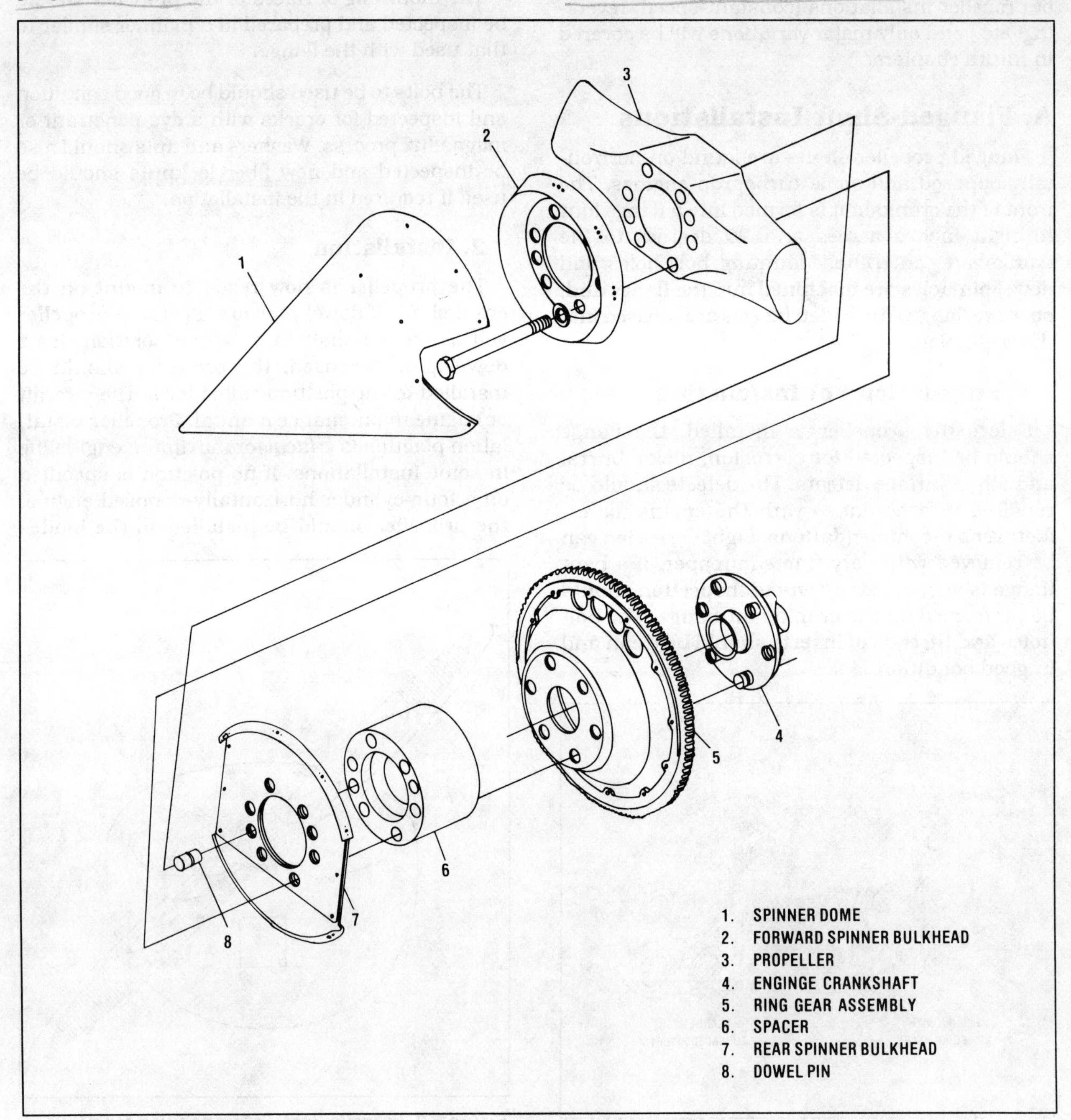

Figure 5-3. Typical flanged-shaft installation with spinner and space.

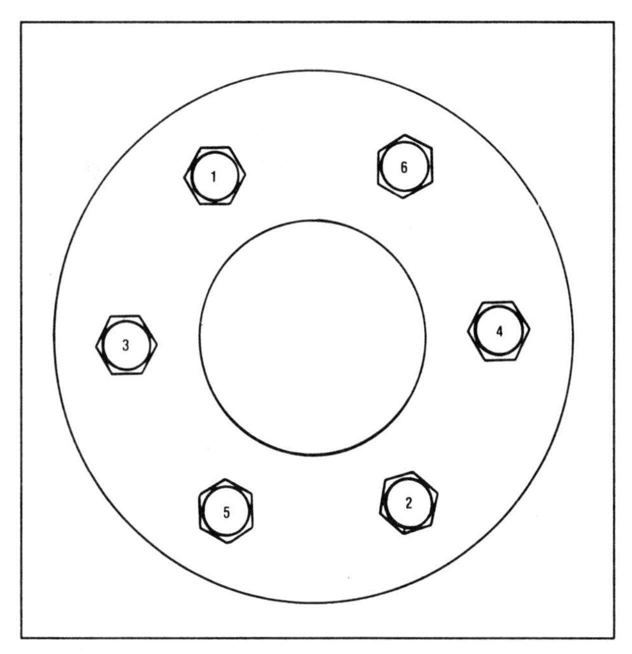

Figure 5-4. Installation torque sequence.

distributes the compression load of the bolts over the surface of the boss.

Spacers between the propeller and flange are installed in accordance with the aircraft manufacturer's manual.

Once the bolts are installed and properly torqued, the propeller is tracked and safetied as discussed later in this chapter.

B. Tapered-Shaft Installations

Tapered-shaft crankshafts are found on older model horizontally-opposed engines of low horsepower. This style of crankshaft requires the use of a hub to adapt the propeller for mounting on the shaft.

1. Pre-Installation Checks

Before the propeller is installed on the crankshaft, the shaft should be inspected carefully for corrosion, thread condition, cracks and wear in

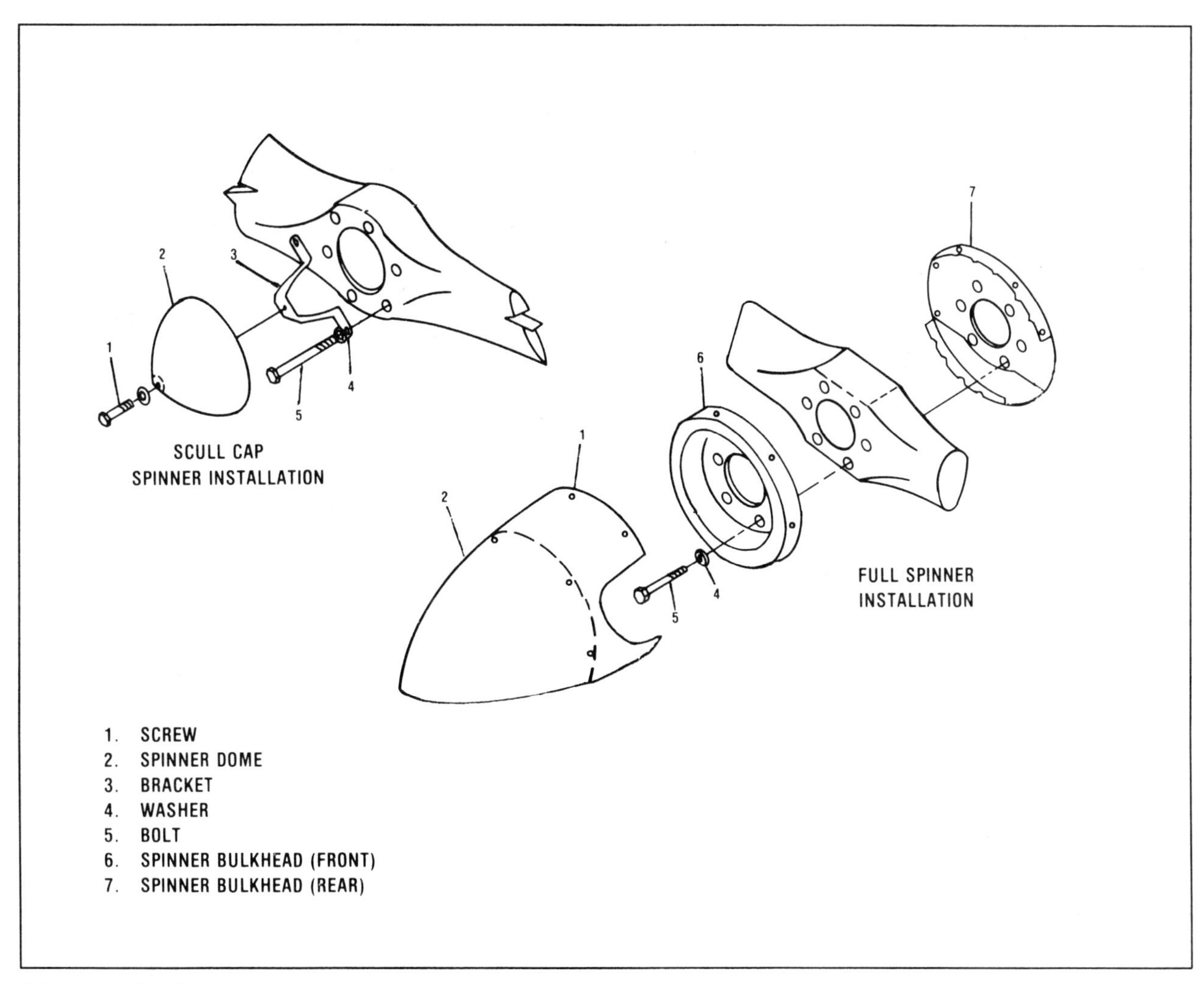

Figure 5-5. Typical spinner installations.

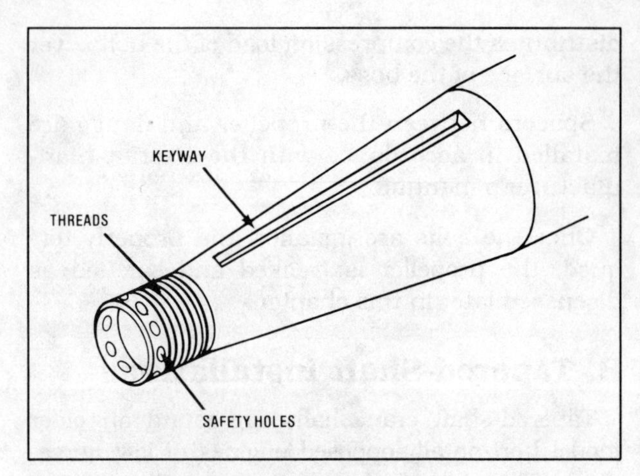

Figure 5-6. Tapered crankshaft.

the area of the keyway. The keyway inspection is critical as cracks can develop in the corners of the keyway and result in the crankshaft's breaking. A dye-penetrant inspection of the keyway area is advisable at each 100-hour and annual inspection, and each time the propeller is removed.

If surface irregularities are found, dress or polish out the defects as the engine manufacturer recommends.

The hub components and mounting hardware should be inspected for wear, cracks, corrosion, and warpage. Correct defects as necessary. A dye-penetrant or magnetic inspection of the hub and bolts is recommended.

The fit of the hub on the crankshaft should be checked by the use of a *liquid transfer ink* such as Prussian Blue. The Prussian Blue is applied in a thin, even coating on the tapered area of the

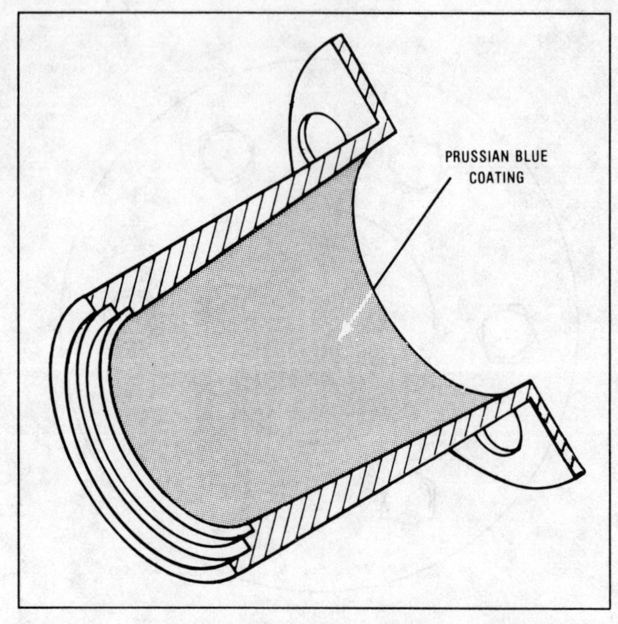

Figure 5-8. Prussian Blue transfer on hub cross-section.

crankshaft. With the key installed in the keyway, the hub is then installed on the shaft and the retaining nut is tightened to the installation torque.

The hub is then removed and the amount of ink transferred from the crankshaft to the hub is noted. The ink transfer should indicate a minimum contact area of 70%. If less than 70% contact area is indicated, the hub and crankshaft should be checked for surface irregularities such as dirt, wear, and corrosion. The surfaces may be lapped to fit by removing the key from the crankshaft and lapping the hub to the crankshaft with a polishing compound until a minimum of 70% contact area is achieved.

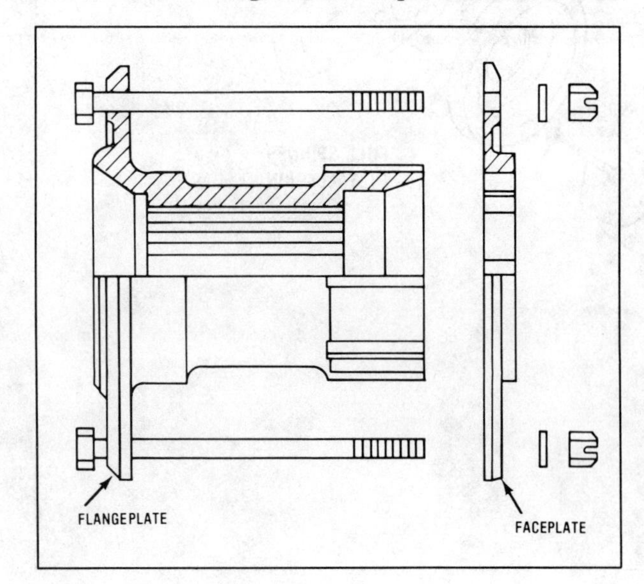

Figure 5-7. Propeller hub.

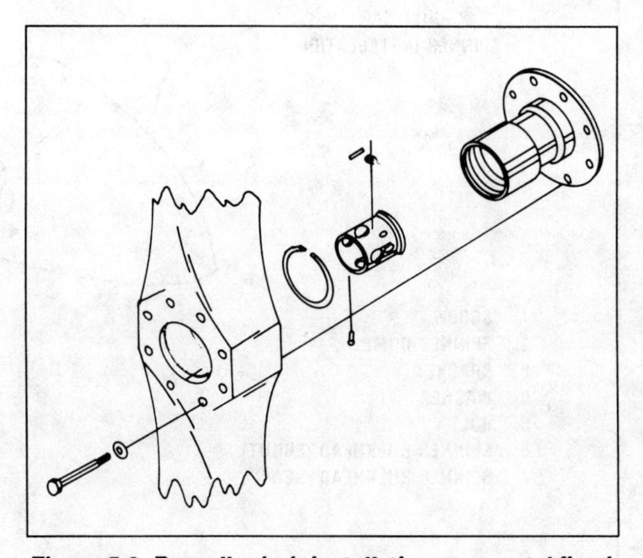

Figure 5-9. Propeller hub installation on a metal fixed-pitch propeller — exploded view.

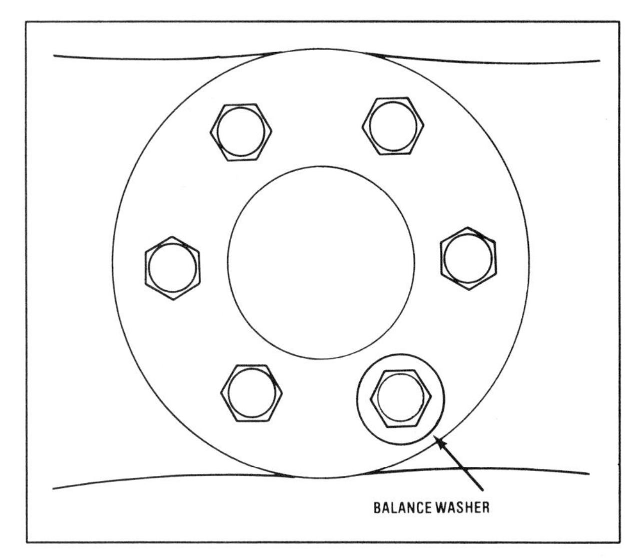

Figure 5-10. Balance washer is used to final balance the propeller-hub assembly.

This inspection and corrective action may be done with the propeller installed on the hub, if desired. Once sufficient contact is obtained, the hub and shaft are cleaned of ink and polishing compound.

Place the propeller on the hub and, if a wood propeller is used, position the faceplate on the front of the boss. Make sure that the propeller is installed on the hub with the blades in the correct position in relation to the keyway if specified by the engine manual. Install the bolts, washers, and nuts as required to mount the propeller on the hub. Tighten the bolts in the same manner as for a flanged-shaft installation. Check the propeller balance with the hub installed and correct the balance by placing approved balancing washers under the bolt head or nut at the light position on the hub.

2. Installation

Apply oil or anti-seize to the crankshaft, making sure that the key is installed properly, and place the propeller-hub assembly on the shaft. Install the retaining nut and torque the nut to the proper value. Install the ***snap ring***, track, and safety the propeller. (Refer to latter portion of this chapter for a discussion of tracking and safetying.)

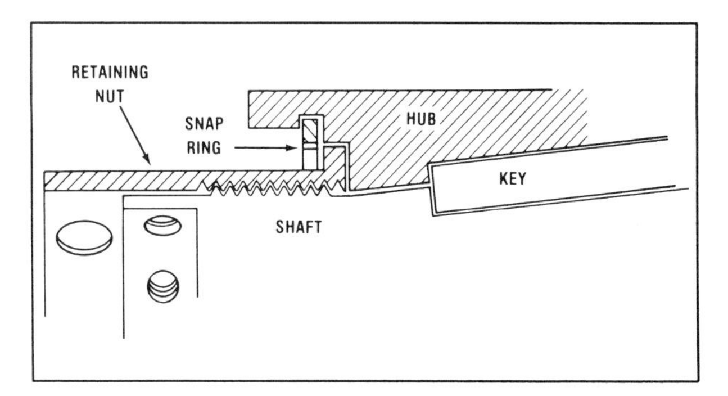

Figure 5-11. Cross section of snap ring installation.

3. Removal

To remove the propeller from the tapered shaft, remove the safety, back the retaining nut off with a bar to pull the propeller from the shaft. A snap ring is required so that the retaining nut can be connected to the hub and pull the hub off the shaft as the nut is unscrewed. If the snap ring is not installed, hub removal may be very difficult.

C. Splined-Shaft Installations

Splined crankshafts are found on most radial engines and some opposed, in-line, and turboprop engines. The splined shaft is characterized by splines and voids of equal dimensions and on many engines a master spline formed between two splines so that a hub will fit on the shaft in only one position.

1. Pre-Installation Checks

Inspect the crankshaft for cracks, surface defects, and corrosion. Repair defects in accordance with the engine manufacturer's directions.

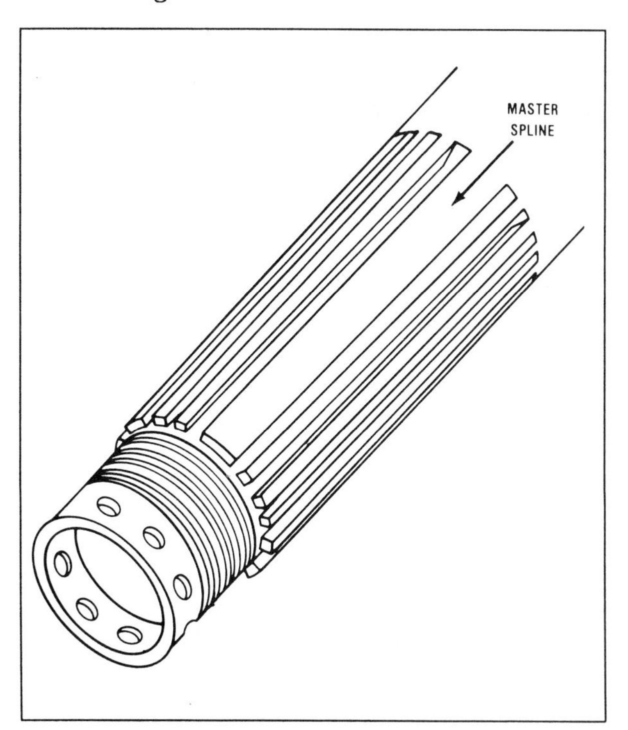

Figure 5-12. Splined crankshaft.

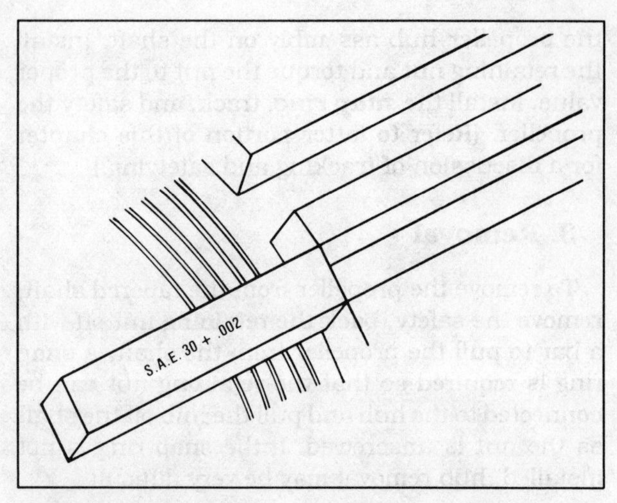

Figure 5-13. Checking for spline wear with a go no-go gauge.

The splines on the crankshaft and on the hub should be inspected for wear by the use of a ***go no-go gauge***. The gauge is 0.002 inches larger than the maximum space allowed between the splines. The crankshaft or hub is serviceable if the gauge cannot be inserted between the splines for more than 20% of the spline length. If the 20% value is exceeded, the hub or crankshaft is worn excessively and should be replaced.

The hub and bolts should be inspected in the same manner as for a tapered-shaft installation.

Cones are used to center the hub on the crankshaft and should be inspected for general condition. The rear cone is made of bronze with a cut at one place to allow flexibility during installation and assure a tight fit when installed. The front cone is in two pieces that are a matched set and must be used together. The front cones are made of steel and are marked with a serial number to identify the mates in a set.

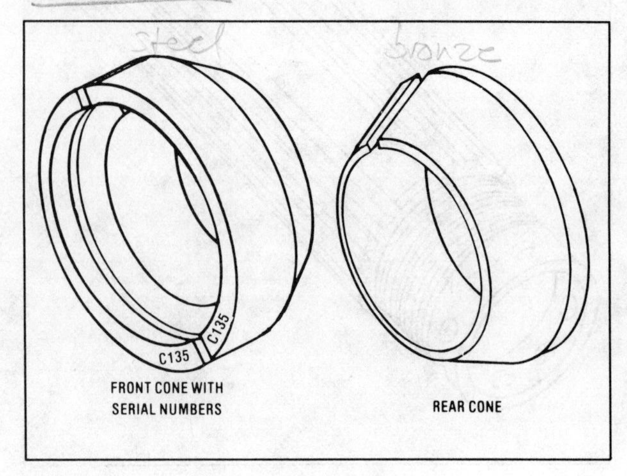

Figure 5-14. Front and rear cones.

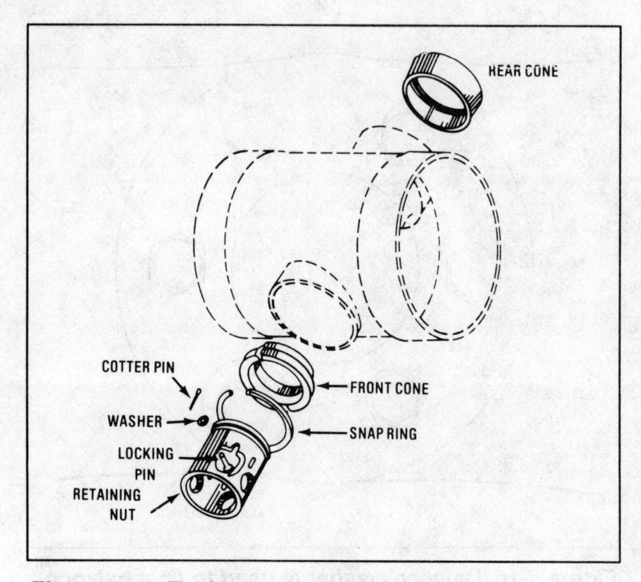

Figure 5-15. Typical splined-shaft installation.

If the front cone is new, the halves will be joined together and will have to be separated with a hacksaw. After the halves are separated, the cut surfaces will have to be filed and polished smooth. In addition, they may have to be marked with an arbitrary serial number by the use of an engraving tool.

2. Trial Installation

The rear cone and, in some installations, a bronze spacer, is place on the crankshaft and pushed all the way back on the shaft. A coat of Prussian Blue is applied to the rear cone. The hub

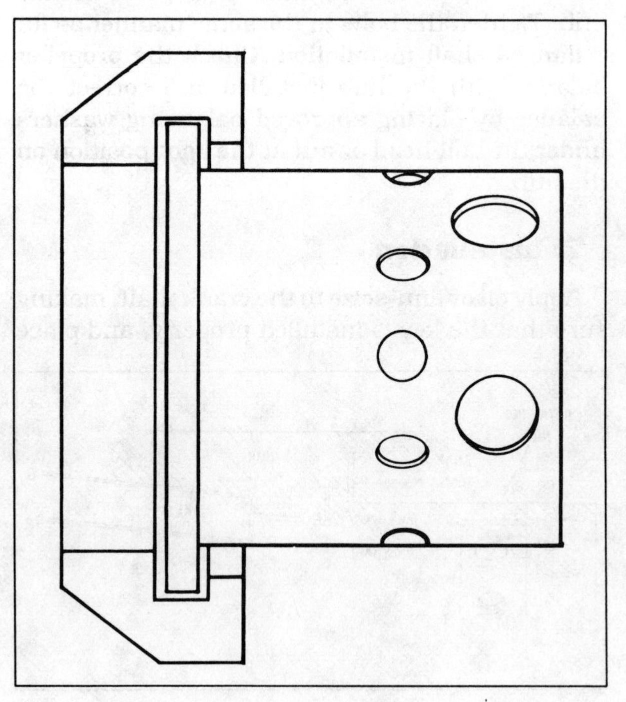

Figure 5-16. Front cone half installed on retaining nut.

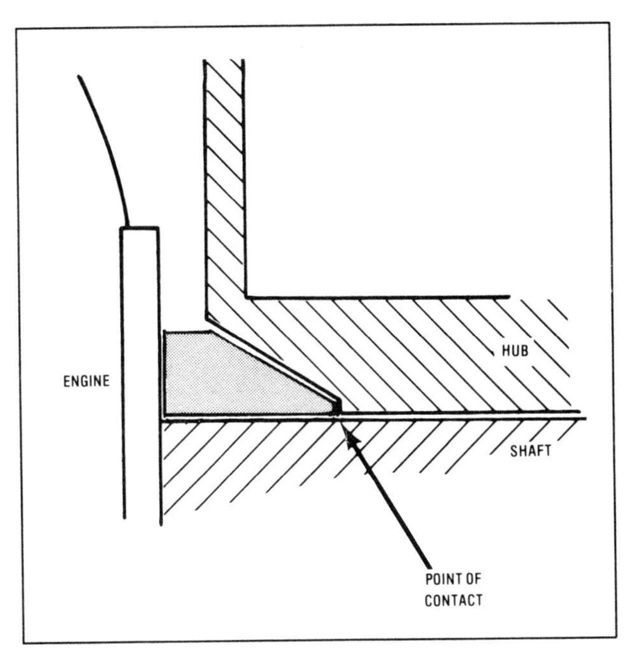

Figure 5-17. Rear cone bottoming.

is then placed on the shaft, with care taken to align the hub on the master spline, and the hub is pushed against the rear cone. The front cones are placed around the lip on the retaining nut, coated with Prussian Blue, installed in the hub, and the nut is tightened to the proper torque.

The retaining nut and front cone are removed and the amount of Prussian Blue transferred to the hub is noted. A minimum of 70% contact is required. The hub is then pulled from the crankshaft and the transfer from the rear cone is checked. Again, a minimum of 70% contact is required. If contact is insufficient, the hub can be lapped to the cones by the use of a special lapping fixture.

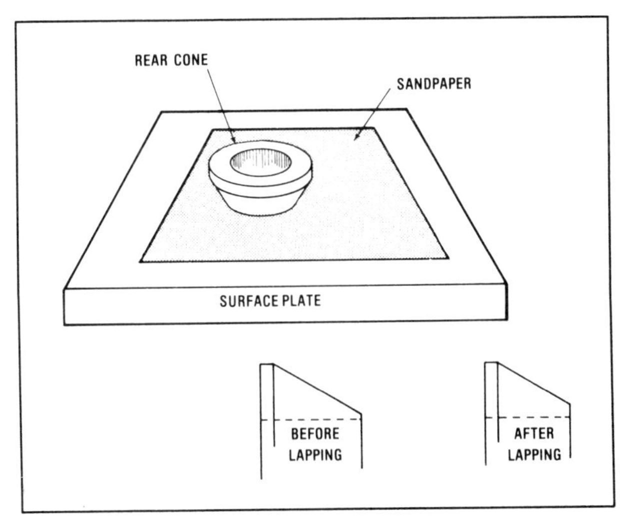

Figure 5-18. Using a surface plate and sandpaper to lap the apex of a rear cone.

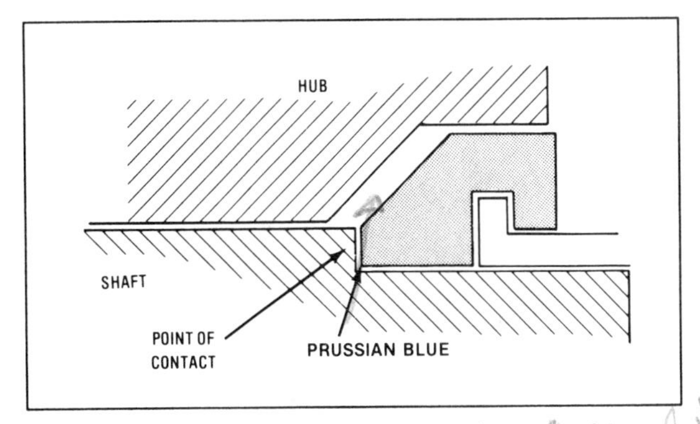

Figure 5-19. Front cone bottoming.

If no transfer to the rear cone occurs during the transfer check, a condition known as rear cone bottoming exists. This happens when the apex, or point, of the rear cone contacts the land on the rear seat of the hub before the hub can seat on the rear cone. Rear cone bottoming is corrected by removing up to 1/16 of an inch from the apex of the cone. Sandpaper placed on the surface plate may be used to assure an even removal of metal.

Front cone bottoming occurs when the front cone bottoms on the splines of the crankshaft before contacting the seat on the hub. Front cone bottoming is indicated by the hub's being loose on the shaft when the retaining nut is tight and no transfer of Prussian Blue to the front hub seat. Front cone bottoming is corrected by placing a spacer of no more than 1/8-inch thickness behind the rear cone. This moves the hub forward and allows the front cone to seat properly. Some installations require a thick spacer (one inch or more) behind the rear cone to assure proper mounting. The corrections noted above are in addition to the spacers called for in the manufacturer's manual. If front and rear cone bottoming cannot be corrected as stated above, and no reason can be determined for the improper seating, consult the manufacturer.

3. Installation

The propeller is installed on the hub in the same manner as used for a tapered-shaft installation. The position of the propeller on the hub in relation to the master spline is *critical.* Some installations require that one blade align with the master spline while other installations require that the blades be perpendicular to the master spline position. Consult the engine maintenance manual for the requirements of a particular installation.

Once the propeller is mounted on the hub, the crankshaft is coated with oil or an anti-seize compound and the propeller-hub assembly is placed

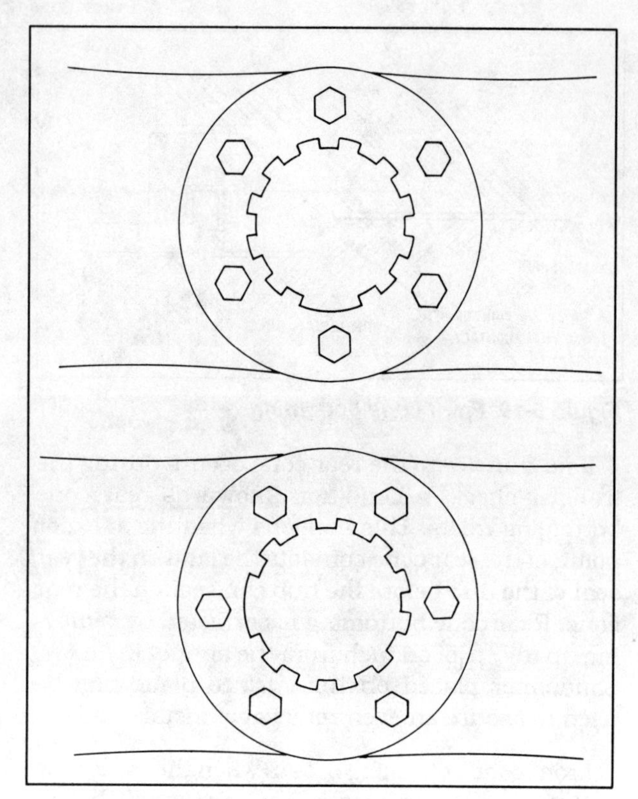

Figure 5-20. Two normal propeller installation positions as related to the master spline.

on the shaft. The retaining nut and front cone are installed and torqued. The snap ring is installed, the ***propeller track*** is checked, and the installation is safetied.

Propeller removal is the same as for a tapered-shaft installation.

D. Tracking The Propeller

Once the propeller is installed and torqued, the track of the propeller should be checked. The track of the propeller is defined as the path which the tips of the blades follow when rotated with the aircraft stationary. For light aircraft with propellers of approximately six feet in diameter, metal propellers can be out of track no more than 1/16 of an inch and the track of a wood propeller may not be out more than 1/8 of an inch.

Before the propeller can be tracked, the aircraft must be made stationary by chocking the wheels so that the aircraft will not move. Next, a fixed reference point must be placed within 1/4 inch of the propeller arc. This may be done by placing a board on blocks under the propeller arc and taping a piece of paper to the board so that the track of each blade can be marked. The propeller is rotated by hand until one blade is pointing down at the paper. (Make sure that the engine is safe as described in Chapter I.) The position of the blade tip is marked on the paper. The propeller is then

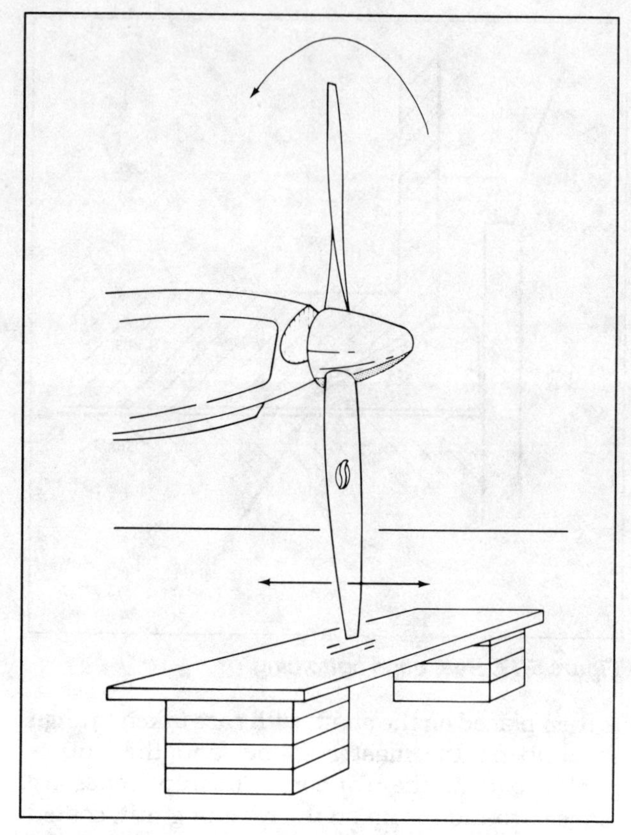

Figure 5-21. Tracking a propeller with a reference board.

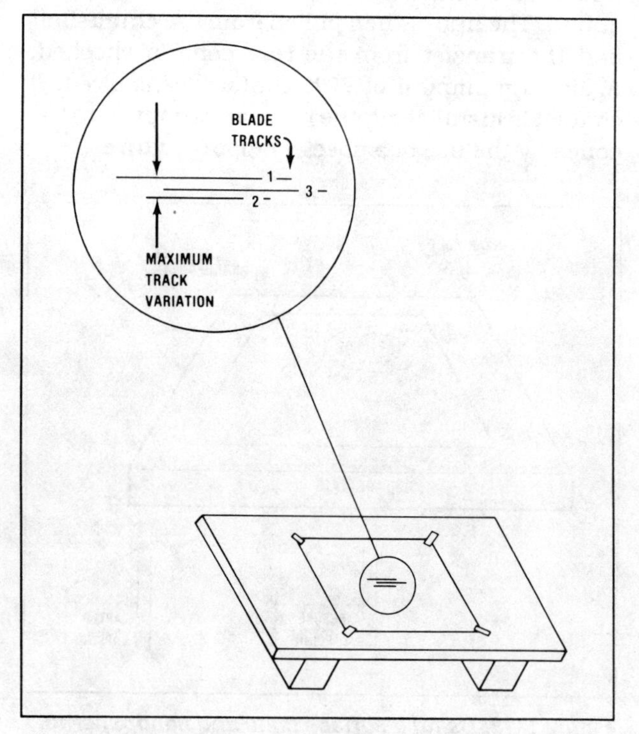

Figure 5-22. Tracking marks for a three-bladed propeller.

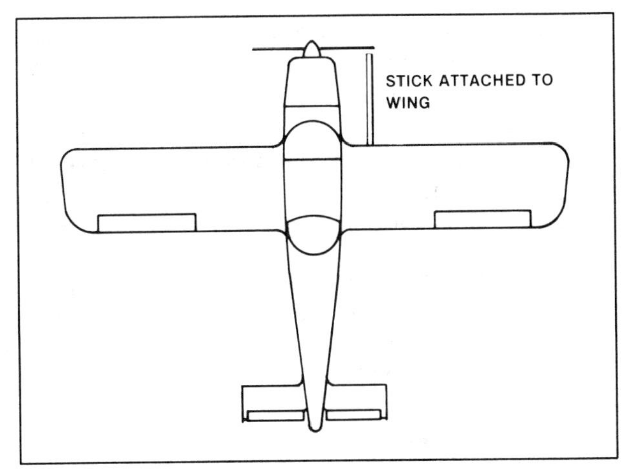

Figure 5-23. Blade track can also be checked with a pointer attached to the airplane.

rotated so that the track of the next blade can be marked on the paper. This is repeated for each blade on the propeller. The maximum difference in track for all the blades should not exceed the limits mentioned for a light aircraft.

If a wooden propeller track is off more than the allowed amount, and presuming that the propeller, hub, and crankshaft are within permissible tolerances, the track may be corrected by the use of shims. A shim shaped to fit halfway around the face of the hub or flange is installed between the flange and propeller on the side with the more rearward tracking blade, so that the blade will be moved forward. The shim should be an approved type of thin brass shim stock of about .002 to .004 inches as necessary to correct the problem. Another shim of the same thickness may be installed between the faceplate and the boss on the side of the more forward blade.

Once the shims are installed, the propeller-hub assembly is reassembled and the propeller is installed on the crankshaft, torqued, and the track is rechecked. If the track is within limits, the installation is safetied. Always consult the manufacturer's manual before using shims as some model propellers do not allow the use of shims.

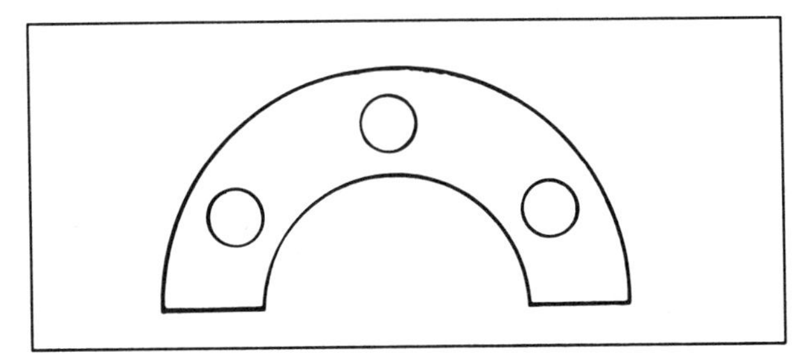

Figure 5-24. A tracking shim.

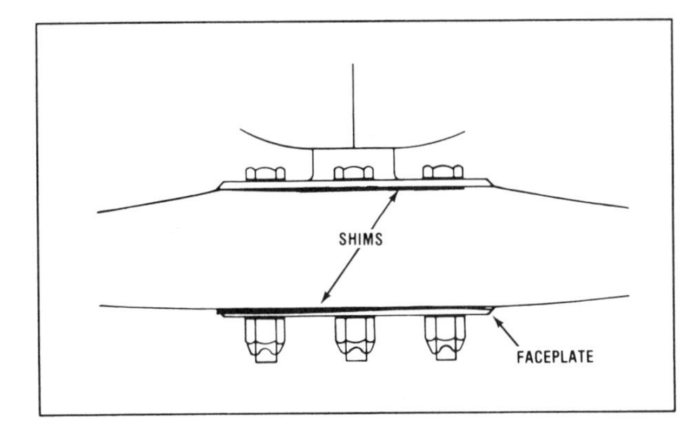

Figure 5-25. Adjusting propeller track on a wood propeller by installing shims.

E. Safetying Propeller Installations

Once a propeller is within track and properly torqued, the installation can be safetied. There is no one way to safety a propeller installation because of the many different types of installations. For this reason only the more commonly used safeties will be discussed here.

A flanged-shaft installation has the largest variety of safety methods because of the many variations in the flange shaft installations. If the flange has threaded inserts installed, the propeller is held on by bolts which screw into the inserts. The bolt heads are drilled and are safetied with 0.041 inch stainless steel safety wire using standard safety wire procedures.

If threaded inserts are not used in a flanged installation, bolts and nuts are used to hold the

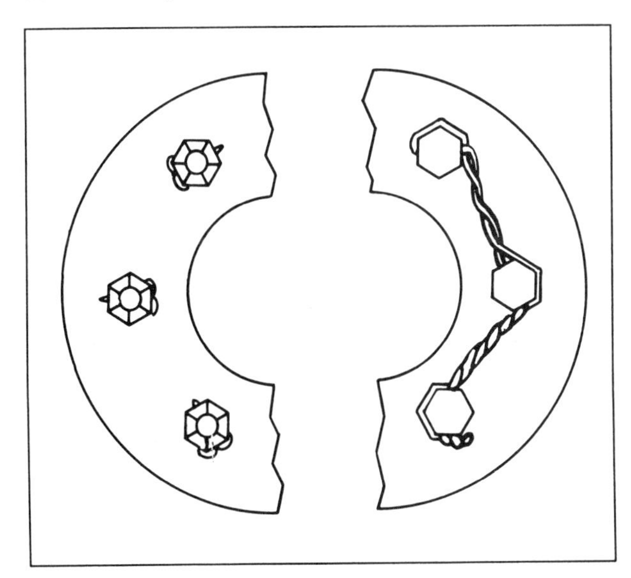

Figure 5-26. Flanged installations may be safetied with safety wire or cotter pins, depending on the style of installation.

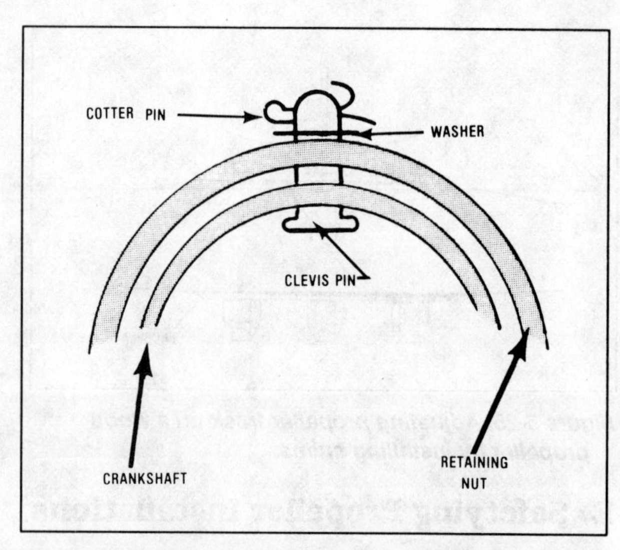

Figure 5-27. Safety of tapered- and splined-shaft installations is done with a clevis pin and a cotter pin.

propeller on the flange. Some installations use fiber lock nuts, which require no safetying, but the nuts should be replaced each time the propeller is removed. Other installations use castellated nuts with bolts drilled in the threaded area and the nuts are safetied to the bolts with cotter pins.

Tapered and splined-shaft installations are safetied in the same way. A clevis pin is installed through the safety holes in the retaining nut and crankshaft. The pin should be positioned with the head toward the center of the crankshaft. A washer is placed over the pin and a cotter pin is installed. Be sure that the head of the clevis pin is toward the center of the shaft so that the centrifugal force will be born by the clevis pin head and not by the cotter pin.

F. Troubleshooting

Troubleshooting the installation of a fixed-pitch usually involves determining the cause of a vibration. When investigating the cause of vibration it is important that the recent history of the propeller be known. Accidents, repairs, and type of operation may give clues to the cause of vibration. Also, structural characteristics of metal propellers may cause a problem to develop that would not be possible in a wood propeller, and visa versa.

Metal propellers may cause vibration if a repair has thrown the propeller nut of balance. The vibration would appear immediately after the repair. The propeller should be rebalanced.

If the propeller is involved in a ground strike, the balance and aerodynamic characteristics may have been altered. If a ground strike is suspected, check with the pilot as not all ground strikes result in readily apparent damage. If a ground strike has occurred, refer the propeller to an overhaul facility for repair.

Some people use the propeller as a handle to pull the aircraft around on the ground. This may result in pulling the blades out of track. The propeller should be overhauled to remove the bend. The same thing may occur if high power settings are used to pull the aircraft out of mud or sand.

If a propeller has recently been repitched, the propeller may tend to return to the original pitch. This would be indicated by an increase in vibration during several hours of operation. The problem cannot be corrected and the propeller should be replaced. This problem often develops in old propellers and is an indication of a breakdown in the metal structure.

Vibration associated with wood propellers is often related to wood damage or moisture in the propeller. If a wood propeller is stored improperly or the aircraft has been idle for a period of time, moisture may be concentrated in one blade, causing the propeller to be out of balance. The moisture will redistribute itself if the propeller is placed in a horizontal position for several days.

Wood propellers may warp, resulting in a change in the aerodynamic characteristics of the blade and causing aerodynamic imbalance. These propellers should be replaced.

Reasons for vibrations that are common to both wood and metal propellers include improper overhaul, uneven torquing of mounting bolts, improper tracking, loose retaining nut, front or rear cone bottoming, and improper installation position in relation to the crankshaft.

If vibrations persist after checking all of the causes listed above, there may be a problem with an interaction between the propeller and the engine cowling. This can be reduced by removing the propeller from the hub or flange and shifting its position one bolt hole. Reinstall the propeller and perform an operational check to determine if the vibration level has decreased. This procedure may be repeated until the position of least vibration is determined. Remember to consult the appropriate manuals to determine if a specific propeller position is called for before attempting this procedure.

QUESTIONS:

1. *What are the two purposes for applying a lubricant to the crankshaft flange before installing the propeller?*
2. *Why is the ten o'clock/four o'clock position suggested when installing a propeller on a four-cylinder horizontally opposed engine?*
3. *What is the purpose of a propeller hub?*
4. *When checking for proper seating with Prussian Blue, what is the minimum amount of contact?*
5. *What is the purpose of the snap ring in a tapered-shaft installation?*
6. *What is the maximum amount of penetration allowed when checking a splined shaft with a go no-go gauge?*
7. *What is the purpose of the cones in a splined-shaft installation?*
8. *What is rear cone bottoming and how is it corrected?*
9. *What is front cone bottoming and how is it corrected?*
10. *What is the maximum allowable out-of-track for a light aircraft metal propeller?*
11. *How may the propeller track be corrected?*
12. *What device is used to safety a splined-shaft installation?*
13. *What is the most likely cause of vibration associated with a wood propeller that has been in storage for a period of time?*

can use this kind of prop for
short runways
power - crop dusting lower angle
cruise - higher angle

If adjust angle on a/c
loosen clamps/bolts and retaining
nut

Chapter VI
Ground-Adjustable Propellers

Ground-adjustable propellers are designed so that their blade angles can be adjusted on the ground to give the desired performance characteristics for various operational conditions (low blade angle for short field takeoffs or high-blade angle for increased cruise speed). The adjustable characteristic also allows one propeller design to be used on aircraft designs with varying performance, but using the same engine model.

Ground-adjustable propellers are often found on older aircraft of low to moderate performance (Stearmans, Wacos, etc.).

A. Propeller Construction

The propeller is designed so that the blades can be rotated in the hub to change the blade angles. The hub is in two halves that must be separated slightly so that the blades can be rotated. The hub is held together with clamps or bolts to prevent the blades from rotating during operation.

The propeller blades may be of either wood, aluminum, or steel construction with the root of the blade having ***shoulders*** machined on it so that the blades will be held in the hub against the centrifugal operating loads.

The hub of the propeller is made of aluminum or steel, with the two halves machined as a matched pair. Grooves in the hub mate with the shoulders on the blades. If steel blades are used, the hub will be held together with bolts. If wood or aluminum blades are used, the hub halves will be held together with bolts or clamp rings.

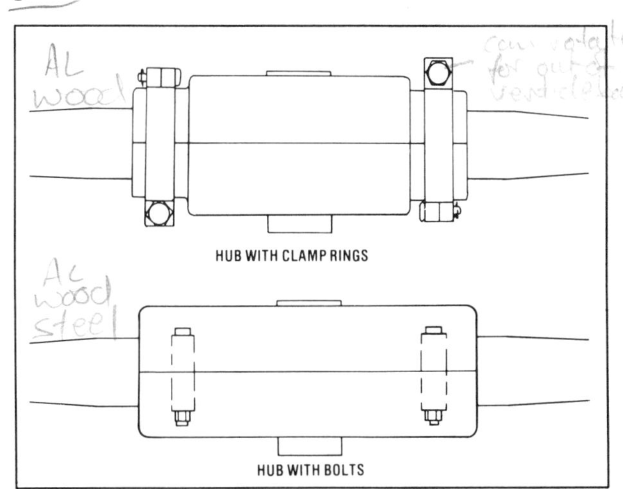

Figure 6-1. Two styles of ground-adjustable propeller hubs.

B. Installation

Ground-adjustable propellers may be designed to fit flanged, tapered, or splined crankshafts. The installation is basically the same as for fixed-pitch propellers.

1. Blade Angle Adjustment

Before the blade angles are adjusted, the reference station must be determined by referring to the propeller or aircraft maintenance manual. The most commonly used reference stations are 30, 36, and 42 inches. Check the aircraft maintenance manual or aircraft specifications to find the blade angle range that may be used for the aircraft. The angle will normally be between seven and 15 degrees. The propeller blade angles can be adjusted on the aircraft or on a propeller bench.

Before the retaining bolts or clamps are loosened, the relative position of the hub and blades should be marked with a red lead, white lead, or grease pencil. *Do not use a graphite pencil!* This marking will allow the change in blade position to be observed and aid in the initial movement of all blades toward the new blade angle.

The propeller is placed in a horizontal position. The hub bolts or clamps and, if the propeller is on the engine, the retaining nut are now loosened

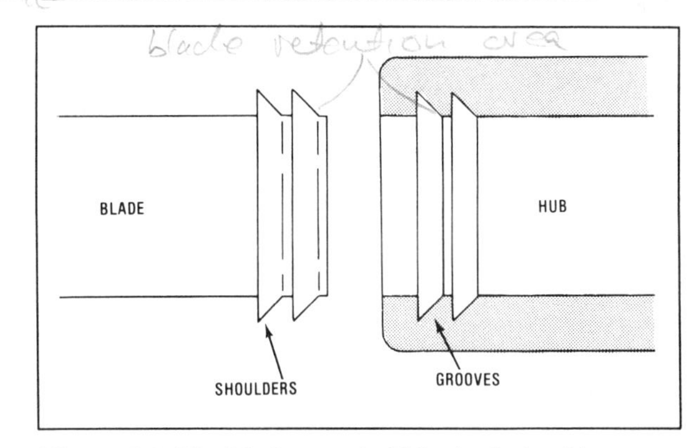

Figure 6-2. The blades are held in the hub with a set of shoulders and grooves.

Figure 6-3. Use a blade paddle to rotate the blades to the desired angle.

until the blades turn freely in the hub. The blades are turned to the desired angle with the aid of propeller paddles as shown in Figure 6-3. It may be necessary to jiggle the blades slightly as they are rotated to prevent binding.

Check the new blade angles with a propeller protractor, then tighten the blade bolts or clamps and the propeller retaining nut. The blade angle may change during the tightening process since the blades tend to hang down when the hub halves are loosened. Determine the amount of blade angle change that occurred during the tightening process, loosen the bolts or clamps and retaining nut, and reset the blade angles, allowing for the blade angle change that will occur during the tightening process. Retighten the hub and measure the blade angles again, they should now be correct.

This procedure may have to be repeated a few times until the blades are at the same angle. Unless otherwise stated in the maintenance manual, an acceptable tolerance for the difference in blade angle between the blades is 0.1 degrees. The blades should be within 0.1 degrees of the desired blade angle.

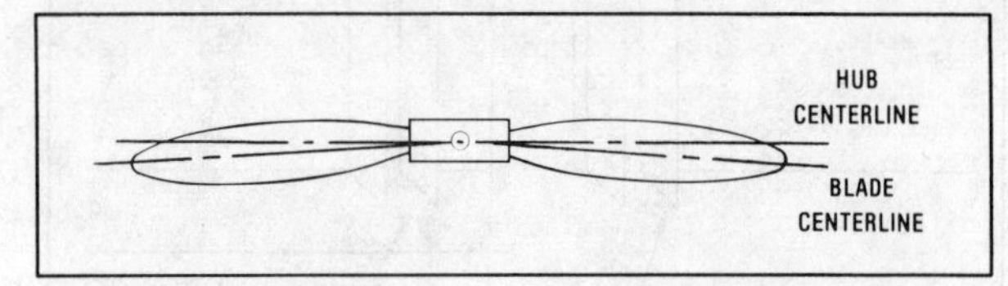

Figure 6-4. Blades tend to hang down when clamps and retaining nuts are loosened while the propeller is on the engine. This must be allowed for when adjusting blade angles.

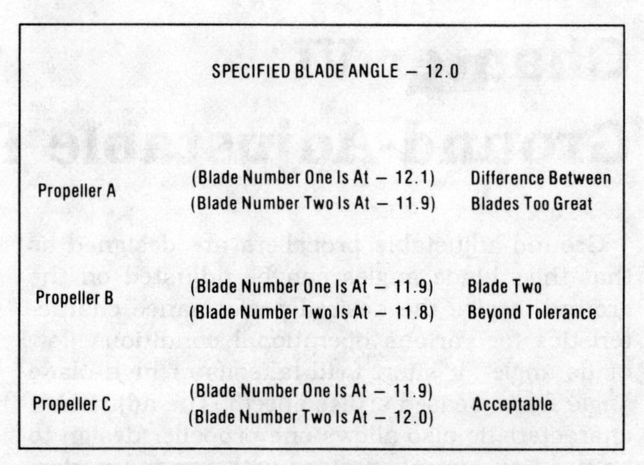

SPECIFIED BLADE ANGLE — 12.0

Propeller A	(Blade Number One Is At — 12.1) (Blade Number Two Is At — 11.9)	Difference Between Blades Too Great
Propeller B	(Blade Number One Is At — 11.9) (Blade Number Two Is At — 11.8)	Blade Two Beyond Tolerance
Propeller C	(Blade Number One Is At — 11.9) (Blade Number Two Is At — 12.0)	Acceptable

Figure 6-5. Examples of acceptable and unacceptable blade angle combinations.

The installation is now torqued, tracked, safetied, and all reference markings are removed.

C. Inspection, Maintenance And Repair

Inspection of the propeller blades is the same as for any propeller, whether it is made of wood or metal. Special attention should be given to the area of the shank where the metal sleeve is used on wood blades and in the area of the retention shoulders and grooves on all blades. A dye- penetrant inspection of these areas is recommended whenever the propeller is disassembled for shipment, local inspection, or repair.

The hub should be inspected closely in the blade retention areas. A dye penetrant inspection is recommended on the external surfaces in this area during routine, 100-hour, and annual inspections. Disassembly of the propeller at these inspection intervals is not normally recommended as more damage may occur during disassembly and reassembly than initially existed.

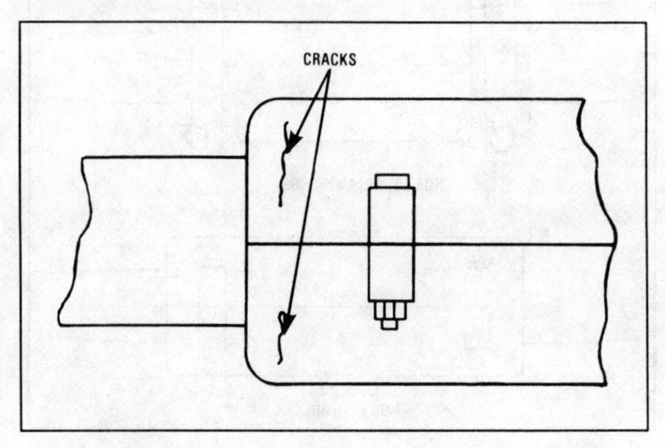

Figure 6-6. Use dye penetrant to check for cracks in the blade retention area.

During 100-hour and annual inspections all torques and safeties should be checked. All nuts, bolts, clevis pins, etc., should be checked for condition and replaced as necessary. Other than cleaning and the repair of defects as covered in Chapter IV, no other routine maintenance need be performed.

The propeller should be overhauled by a repair station at each engine overhaul or as specified by the manufacturer.

D. Troubleshooting

The determination of operational problems with a ground-adjustable propeller usually consists of determining the cause of excessive vibration.

The following are causes of vibration and the solutions are evident by correcting the problem in a manner previously discussed: different blade angles set on each blade; loose retaining nut; blades loose in the hub; front cone bottoming; rear cone bottoming; excessive hub or crankshaft spline wear; insufficient cone or tapered shaft contact area.

Vibration may also be caused by the propeller being out of track. This problem will require that the propeller be sent to a repair station.

Another reason for vibration from a new or recently overhauled propeller, is that one blade has a different length, pitch distribution, weight, or airfoil shape than the other blade. This should be corrected by the manufacturer or an overhaul facility.

QUESTIONS:

1. *How are the blades of a ground-adjustable propeller retained in the hub?*
2. *What devices are used to hold the hub halves together if steel blades are used?*
3. *Which devices must be loosened before the blade angles can be adjusted?*
4. *What types of pencil may be used to mark the hub and blades?*
5. *If the blade angles are being adjusted with the propeller on the aircraft, why may the angles have to be adjusted two or more times?*
6. *What is an acceptable difference in blade angle between the blades?*
7. *Which areas of the hub should be inspected with dye penetrant?*
8. *When should the propeller be overhauled?*

Chapter VII
Automatic Pitch-Changing Propellers

Automatic pitch-changing ***propellers*** are designed so that the blade angle will change in response to operational forces providing the desirable load on the engine as flight conditions change. At high power setting and low airspeeds, as in a climb, the blade angle will decrease to provide good climb performance. At high airspeeds and moderate power settings the blade angle will increase to provide good cruise performance. There is no cockpit control to cause this change, *the propeller is an independent and automatic unit.*

Automatic pitch-changing propellers are not now in wide usage due to the controllability achieved with modern constant-speed systems. However, this propeller design is still found on some older light aircraft such as Swifts, Bellancas, and early Cessnas. The design most commonly used is the Koppers Aeromatic® propeller, the model which will be discussed in this chapter.

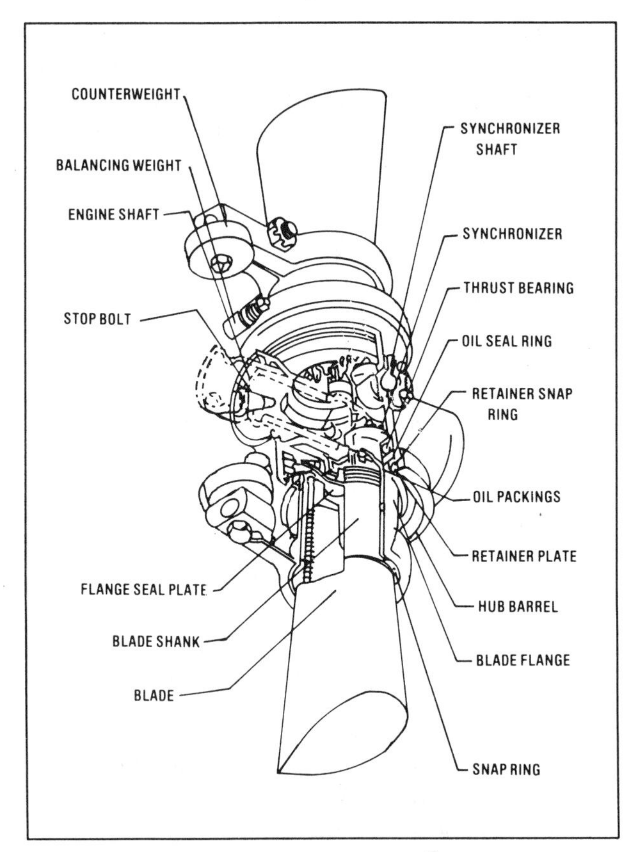

Figure 7-1. Cutaway of an Aeromatic® propeller.

Although not widely used, this style of propeller is covered because its principles of operation are similar to those used on more modern constant-speed designs and it is not complicated by the use of cockpit controls and governors.

A. Theory Of Operation

The Aeromatic® propeller uses the natural forces acting on the blades and the counterweights to automatically achieve the desired change in blade angle for different flight conditions. Automatic operation represents a balance between the forces which tend to increase blade angle and those which tend to decrease blade angle.

The interaction of the forces which determine the blade angle may become confusing, so it is suggested that the student refer to the accompanying illustrations while studying this section. Also, note that the centerline of the propeller blade is *behind* the centerline of the hub and the angle between the hub centerline and the blade centerline remains constant as the blade angle changes. This arrangement results in the blade moving forward of the hub plane of rotation when at low angles and moving behind the hub plane of rotation when at high blade angles.

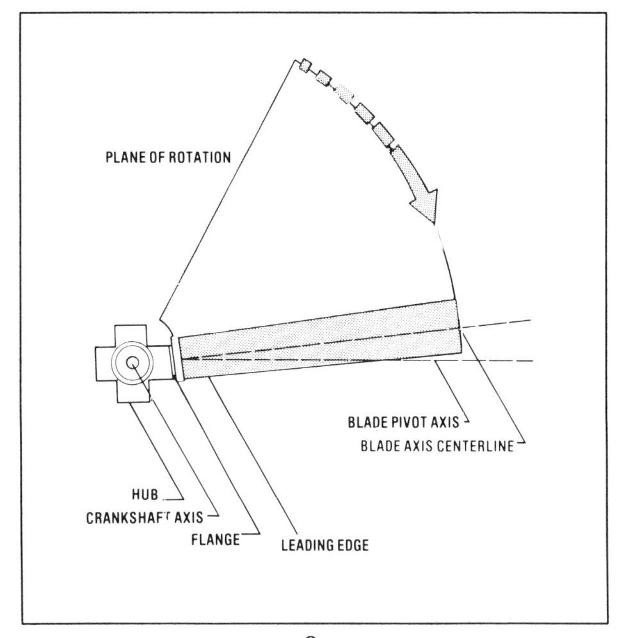

Figure 7-2. The Aeromatic® blade centerline lags behind the hub centerline.

Consider the aircraft to be in cruising flight. When the throttle is pushed in, engine horsepower increases and causes the RPM to increase. An increase in RPM will result in an increase in centrifugal twisting moment which tends to decrease blade angle. The increase in lift on the propeller caused by the RPM rise will cause the blade to move forward and help to decrease the blade angle. The higher RPM results in a higher blade angle of attack with the result that the center of lift on the blade moves toward the leading edge to further increase the movement toward a lower blade angle.

As the aircraft starts to accelerate in response to the higher power setting, the blade angle of attack starts to decrease and the center of lift on the blade starts moving rearward. The initial increase in RPM caused an increase in the force of the counterweights toward a higher blade angle, but this was insufficient to overcome the combined forces moving the blade toward a lower blade angle. However, as the center of lift moves rearward, tending to increase blade angle, and the system RPM increases with the rise in airspeed, the force generated by the counterweights becomes sufficient to cause an increase in blade angle.

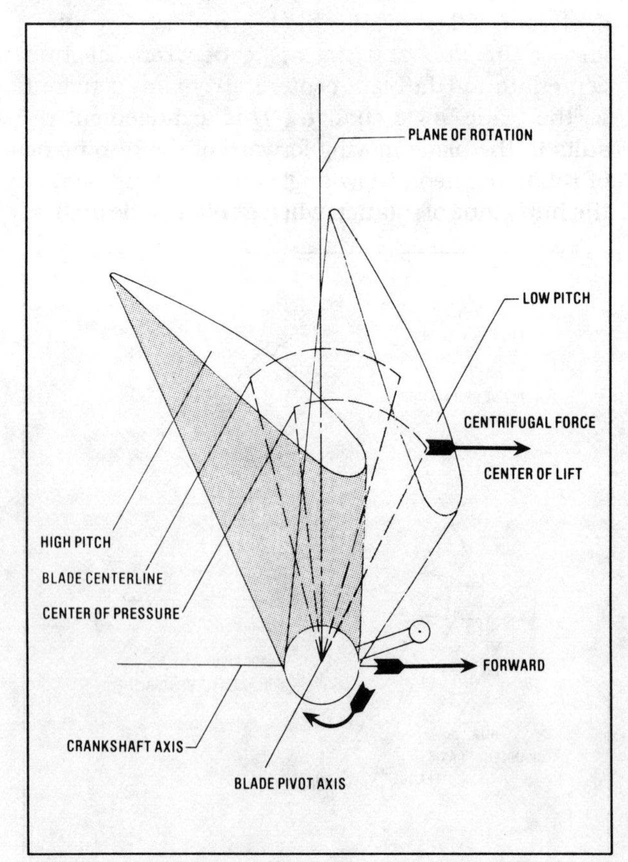

Figure 7-3. Aeromatic® propeller forces which decrease blade angle.

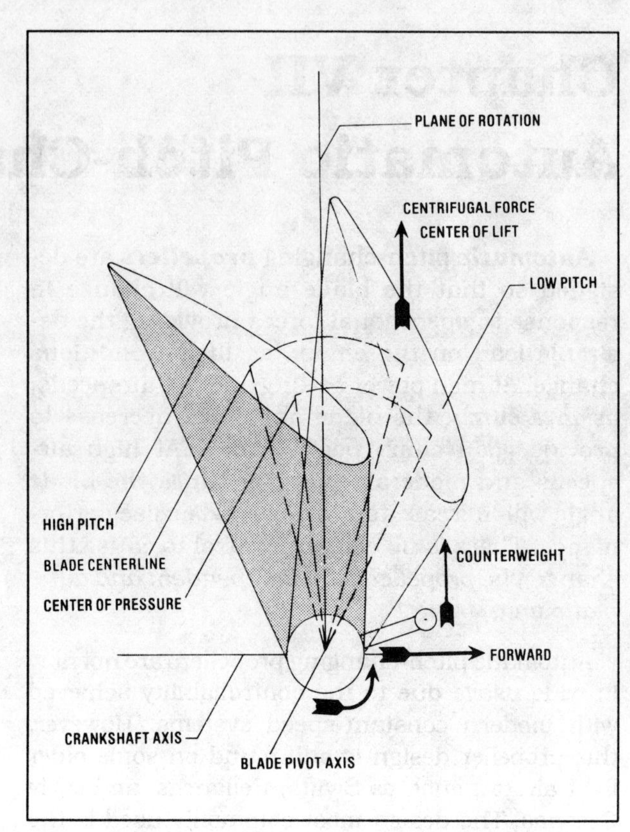

Figure 7-4. Aeromatic® propeller forces which increase blade angle.

At the point where the airspeed no longer increases with the higher power setting, the opposing forces balance out and the blades assume the most desirable angle for the existing conditions of airspeed and engine power output.

B. Propeller Construction

The Aeromatic® propeller blades are made of thin wood laminations, usually covered with a black plastic coating for surface protection, and have a metal leading edge. The shank of each blade is mounted in a steel sleeve and is attached to the hub. A segmented gear is mounted on the butt of the blade and a counterweight bracket is located on the outside of the sleeve.

The hub may be designed to fit any of the crankshaft styles previously discussed. The hub is made of steel and contains all of the bearing surfaces necessary for smooth propeller operation.

A segmented synchronizer gear is located inside the hub and meshes with the gears on the butt of each blade so that each blade turns the same amount during operation, thereby maintaining the same blade angle on each blade at all times.

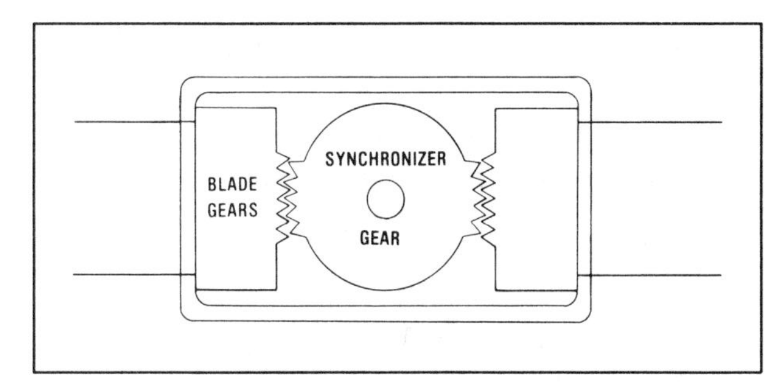

Figure 7-5. A synchronizer gear is used to keep the blades at the same angle.

There are four stop bolts which extend into the hub and are used to set the high and low blade angles of the propeller blades. These bolts contact stop lugs which are part of the blade butt structure.

The counterweights on the blade sleeves extend forward of the propeller and carry a series of weights with different size weights which may be added or removed as necessary to adjust the operation of the propeller. The weights are held on the counterweight bracket by a bolt and castellated nut.

C. Installation

The installation of an Aeromatic® propeller is the same as for a fixed-pitch propeller. When checking the track, be sure that the blade is at full low or high blade angle. Remember, the blade path moves forward as blade angle decreases.

D. Inspection, Maintenance And Repair

If a natural wood finish is used on the blades, inspect and repair the blades as described in Chapter IV. If a plastic coating is used, inspect the metal tipping as in Chapter IV. The plastic coating should be inspected for abrasion, cracks, and missing sections. The blade should be twisted and flexed slightly at the tip to check for broken wood sections under the coating, indicated by wrinkles appearing in the coating.

Small surface defects in the plastic coating can be repaired by the use of a field repair kit available from the propeller manufacturer. Large defects in the coating or an indication of a crack in the wood should be referred to a repair station, as repair of these defects constitutes a major repair. Temporary repairs to small defects in the plastic coating that expose the wood can be done by applying two coats of clear nitrate dope to the area, allowing one hour of drying time between coats. This is only a temporary repair and the permanent repair should be performed as soon as possible.

The hub and blade sleeves should be inspected for surface defects and cracks with a magnifying glass and dye penetrant. Any defects found should be referred to an overhaul facility for correction.

The Aeromatic® propeller is lubricated by a self-contained oil supply which should be checked at 100-hour and annual inspections. The oil level is checked by turning the propeller 45 degrees from the horizontal with the hub oil plug up. Remove the plug and check to see that the oil level is level with the opening. As necessary, fill the hub with the oil grade called for in the aircraft or propeller maintenance manual. If the oil level is too high, erratic or sluggish operation may result. Oil leakage around the blades or hub indicates defective oil seals and the propeller must be sent to an overhaul facility to replace the seals.

When the propeller is placed in storage, the hub should be completely filled with oil and placed in a horizontal position, as is standard for all wood propellers. The propeller should be turned once every two weeks to keep a coating of oil on all the internal parts and prevent corrosion.

Aeromatic® propeller blades should move freely and easily when turned by hand. If this does not happen, defective blade bearings are indicated and the propeller should be overhauled. If there is excessive movement of one blade while the other blade is held rigid, excessive wear has occurred on the blade gear segments or the synchronizer gear is worn. Again, have the propeller overhauled.

E. Troubleshooting

If vibration is noticed during operation, the blades should be checked for gear wear as mentioned above. The vibration may also be caused by the propeller being out of track (1/8-inch maximum difference between the blades) or the propeller being out of balance. If the track is beyond the limit, or the balance cannot be corrected by placing the propeller horizontal to allow even moisture distribution, send the propeller to an overhaul facility for correction.

If the RPM is too high for flight conditions, the weights on the counterweights may not be sufficient. Refer to the propeller and aircraft manual and add washers to the counterweights as necessary. The opposite holds true if the RPM is too low, remove washers as necessary. Make sure that an equal amount of weight is added or removed from each counterweight. When test flying the aircraft during counterweight adjustment, make sure that the spinner is installed if used on the aircraft. The propeller may have slightly different counterweight and aerodynamic action depending on whether or not the spinner is installed.

QUESTIONS:

1. *Name two forces that will cause an increase in blade angle on an Aeromatic® propeller.*
   CENTRIFUGAL FORCE + AERODYNAMIC LIFT ON BLADE
2. *What two factors influence blade angle of attack?* RPM + A/C AIRSPEED
3. *As the aircraft transitions from climb to cruise with the same throttle setting, will blade angle increase or decrease?*
   INCREASE
4. *What is the purpose of the segmented synchronizer gear?* EACH BLADE TURNS SAME AMOUNT
5. *What temporary repair may be made for a small crack in the plastic coating?*
   TWO COATS OF CLEAR NITRATE DOPE
6. *What would be the result if the propeller were serviced with too much oil?* ERRATIC + SLUGGISH OUT.
7. *What preparations are necessary before placing the propeller in storage?* FILL HUB WITH OIL
8. *If the operational RPM is too high, are weights added to or removed from the counterweights?*
   ADD
9. *Why must the propeller spinner be installed when flight testing the propeller for RPM adjustment?* AIRFLOW OVER COUNTERWEIGHTS MAY AFFECT BLADE ∠
10. *What should be done with the propeller if the track is out by more than 1/8-inch?*
    SHIMS ON A FLANGE INSTALLATION OR PROP OVERHAUL

Chapter VIII
Controllable-Pitch Propellers

A controllable-pitch propeller is designed so that the pilot can select any blade angle within the propeller's range regardless of the aircraft operational conditions. This type of propeller is occasionally found on aircraft of the post-World War II era and was eventually refined into the Beechcraft electric constant-speed propeller. This propeller design allowed light aircraft with as little as 65 horsepower to have the advantages of a variable-pitch propeller without the complexity and expense of a constant-speed system. The most popular design is known as a Beech-Roby propeller.

A. System Components

1. Propeller Construction

The blades of Beech-Roby propellers are of wood construction and may have a varnished finish or be coated with plastic. The blades are in a metal shank as in the Aeromatic® and ground adjustable designs.

The hub may be a one-piece shell or composed of two halves which are held together with bolts. Internal components consist of ball bearings to carry the operational loads between the hub and blades and the pitch-changing mechanism as discussed in System Operation.

The propeller may be designed to mount on any of the crankshaft styles covered in Chapter V.

The range of propeller travel is set by mechanical stops on the drive gear behind the propeller.

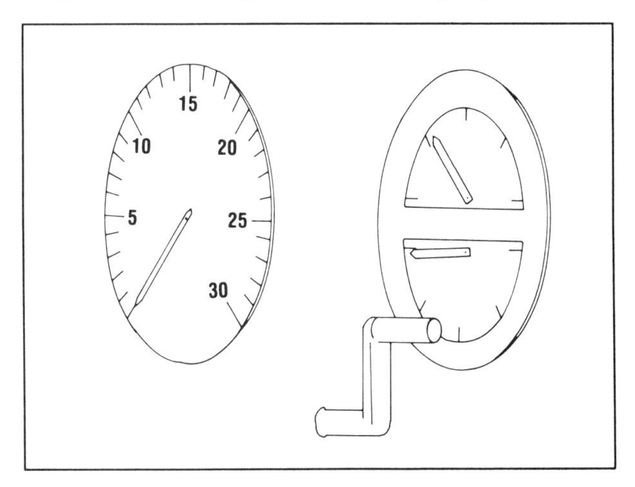

Figure 8-1. Propeller control crank on instrument panel.

2. Cockpit Control

The cockpit control is a crank handle which directly drives the pinion gear through a flexible cable. The cable housing is supported at several points to prevent abrasion and excessive cable flexing.

3. Design Variations

Some more sophisticated designs of the Beech-Roby system use a toggle switch in the cockpit, operating an electric motor to drive the pinion gear. In this design, microswitches are placed with the mechanical stops to limit the blade angle ranges and automatically shut off the electric motor at the maximum and minimum blade angles.

B. Theory Of Operation

The Beech-Roby propeller is operated from the cockpit by a crank handle mounted on the instrument panel. When the crank is turned, a flexible cable (similar to a tachometer drive cable) rotates and turns a pinion gear mounted in a bracket directly behind the propeller. The pinion gear meshes with a drive gear which fits around the crankshaft and is mounted to the engine crankcase. The drive gear can rotate through a limited range on its mount.

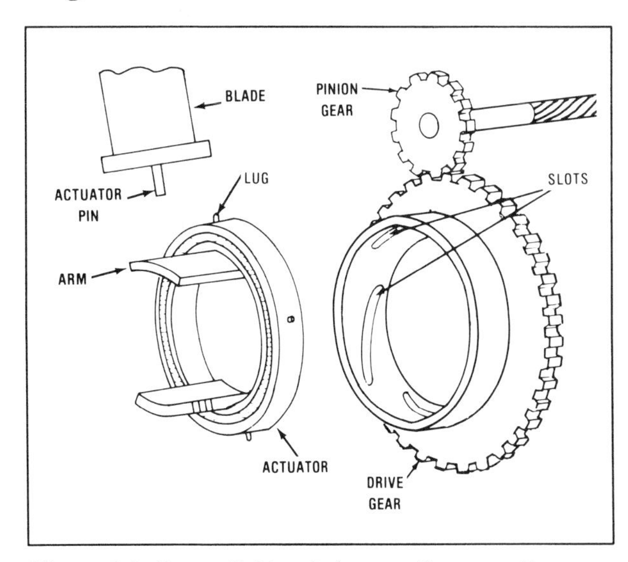

Figure 8-2. Controllable-pitch propeller operating mechanism.

The drive gear is grooved internally with spiral slots which mate with the outer race of the actuator (the outer and inner races of the actuator are connected through ball bearings so the inner race can rotate with the propeller). As the drive gear is rotated by the pinion gear, the actuator moves forward or rearward as the lugs move in the drive gear slots.

The inner race of the actuator rotates with the propeller and incorporates two arms which extend forward into the hub. These arms are connected to an actuating pin in the base of each blade through a set of control fingers. As the outer race of the actuator moves forward or rearward, the inner race moves with it, causing the blade angle to change through the connection between the arm and the actuating pin.

C. Installation

The installation of the propeller on the crankshaft is the same as for installations covered in Chapter V. A mounting bracket is required to install the drive gear on the engine case.

If the aircraft is being modified to accept a Beech-Roby propeller, the cowling may have to be altered to allow clearance for the pinion gear. The instrument panel will require a slight alteration to accept the control crank, and provisions will have to be made for bracing the control cable housing between the cockpit control and the pinion gear.

D. Inspection, Maintenance And Repair

The blades are inspected and maintained as discussed in Chapter IV.

The hub should be inspected for cracks, loose blades, and proper torque and safety of the installation. There should be a minimum of play when the blades are checked for rotational security. If movement is excessive, the propeller should be overhauled.

The cockpit control should operate freely without binding or catching through the full range of travel. The cable should be lubricated at 100-hour and annual inspections and should be replaced if any binding or catching is noticed. The cable mounts and support brackets should be inspected for wear, abrasion, and cracks. Repair mounts and brackets in accordance with accepted aircraft structural repair practices.

There should be a minimum of play between the pinion gear and the drive gear and the gears should be free of dirt and corrosion. The gears should not be lubricated since lubricants will attract dirt and abrasives and cause wear.

If an electric mechanism is used to control the blade angle, inspect and repair the electrical components in accordance with acceptable maintenance practices.

E. Troubleshooting

Standard troubleshooting procedures and corrections apply to controllable-pitch propellers with the following additions.

Vibration may be caused by wear in the actuator lugs, actuator fingers, drive gear slots, or actuating pins. These wear problems should be corrected by an approved overhaul facility.

If the system RPM is too high or too low, check the blade angle stops and adjust as necessary to comply with the aircraft specifications.

QUESTIONS:

1. *What type of cockpit control is used to operate a Beech-Roby propeller?*
2. *How is the blade angle range adjusted on this propeller?*
3. *What operational condition would indicate a need to replace the control cable?*
4. *What is used to lubricate the gears of the pinion gear and drive gear?*

Chapter IX
Two-Position Propeller System

A two-position propeller system is designed so that the pilot can select one of two blade angles in flight. This capability allows the pilot to place the propeller in a low blade angle for takeoff and climb or in a high blade angle for cruise, similar to having a two-speed transmission in a car.

While this system is not currently being used on any production aircraft, it was very popular in the 1930s for high performance civilian and military aircraft. Aircraft presently using this propeller system include antiques and agricultural aircraft. The basic propeller design and pitch changing principles of the Hamilton-Standard two-position propeller are still in use on aircraft as sophisticated as the Beech King Air and the Cessna Conquest.

A. System Components

1. Propeller

The central component of the Hamilton-Standard counterweight propeller is the ***spider***. The spider is designed for installation on a splined crankshaft and incorporates two or three arms on which the blades are mounted.

Aluminum blades are commonly used, although some early models used wood blades. On the butt of each blade is installed a counterweight bracket which is part of the pitch-changing mechanism.

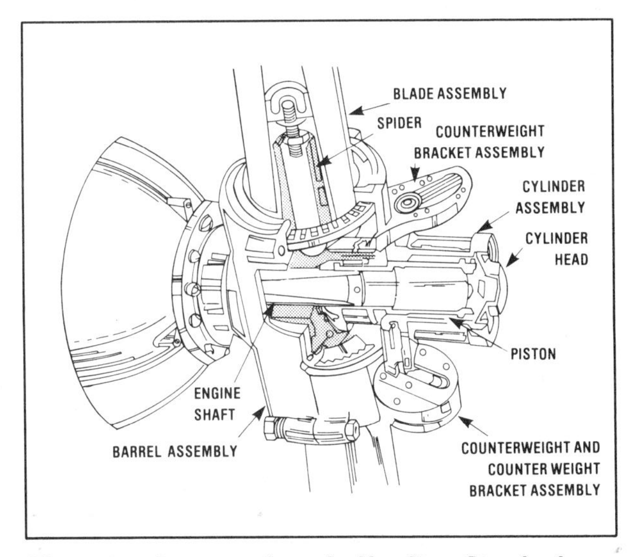

Figure 9-1. Cutaway view of a Hamilton-Standard counterweight propeller.

The blade-bracket assembly is installed on an arm of the spider.

The propeller barrel (hub assembly) is made of two steel halves which are machined as a set and serially numbered. Barrel halves cannot be interchanged between sets. The barrel halves, with required bearings and spacers, are placed around the spider after the blades are installed and are held together with Welch bolts (more on these later).

The cylinder fits between the counterweight brackets and is attached to the brackets by special Allen-head screws which act as follower pins during pitch-changing operations.

The counterweight assemblies are installed on the counterweight brackets. These contain the blade angle setting mechanism of the propeller.

The piston fits through the cylinder and doubles as the propeller retaining nut. To the piston are attached the front cone, the snap ring, and a safety ring. Leather seals are used between the piston and cylinder and are held in place by a seal nut.

If a crankshaft breathing engine is being used, a breather shaft is installed through the center of the propeller to a threaded area in the center of the crankshaft.

The cylinder head and its copper-asbestos gaskets are installed in the forward end of the cylinder and safetied with a locking ring wire.

The Hamilton-Standard propeller designation system can be deciphered in the following manner using a 12D30-235 propeller as an example. The *1* indicates a major modification to the basic propeller design; the *2* indicates the number of blades on the propeller; the *D* indicates the size of the blade shank, based in a Hamilton-Standard designation system; the *30* means that the propeller fits an SAE number 30 splined shaft; the *235* indicates a minor modification of the design; and the *5* in the *235* means that the propeller is designed to rotate clockwise when viewed from the cockpit. If the last number in the minor modification designation is an odd number (1, 3, 5, 7, 9) the propeller rotation is clockwise, and if the number is even (2, 4, 6, 8) the rotation is counterclockwise.

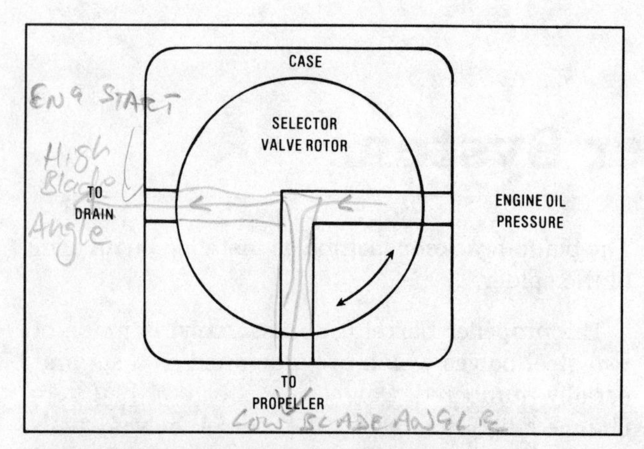

Figure 9-2. Simplified selector valve.

The blade designation for a Hamilton-Standard 6253A-18 propeller blade means: the basic propeller blade design is a *6253;* the blade is a complete assembly with bearing surfaces and other necessary parts, indicated by the *A;* the propeller has been reduced in diameter by 18 inches (each blade is shortened nine inches).

2. Selector Valve

A ***selector valve*** is used to direct oil at engine system pressure to the propeller or drain the oil from the propeller and return it to the engine oil sump. The valve is controlled by the pilot from the cockpit by a propeller control lever. When the propeller control lever is moved forward, the selector valve rotates to direct pressurized oil to the propeller and cause a decrease in blade angle. By moving the control aft, the selector valve rotates to drain oil from the propeller and increase blade angle. The selector valve may be located on the rear accessory case of the engine or on the nose case.

3. Cockpit Control

The cockpit control for the propeller is normally located with the throttle control lever and is connected to the selector valve through flexible cables using pulleys and turnbuckles as necessary for routing and adjusting cable tension. The control lever is moved forward for low blade angle and rearward for high blade angle. There is no intermediate position.

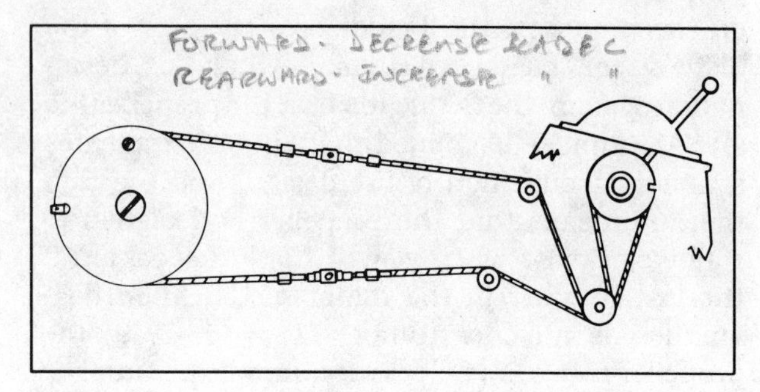

Figure 9-3. An example of a cockpit control mechanism.

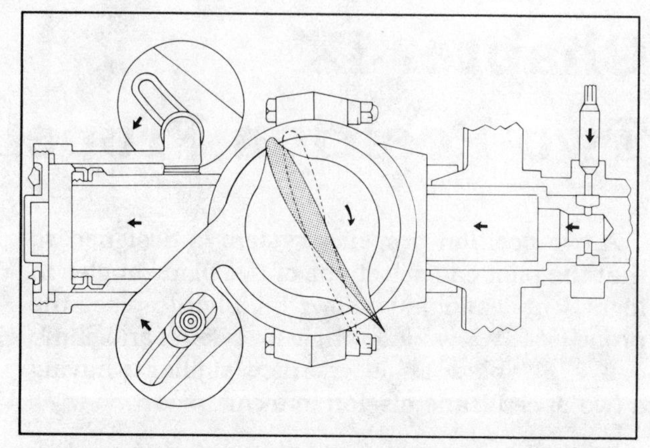

Figure 9-4. Oil pressure moves the cylinder forward and decreases propeller blade angle.

B. System Operation

Two forces are used to cause the blade angle to change — engine oil pressure in the propeller cylinder and centrifugal force acting on the counterweights. The other rotational and operational forces have a minimum of effect on system operation.

When the propeller control lever is moved forward to decrease the blade angle, the selector valve is rotated to direct engine oil pressure (60 to 90 psi) to the propeller cylinder. Oil flows from the selector valve, through passages in the engine nose case, and is delivered to the hollow crankshaft through a transfer bearing.

As the oil pressure moves the cylinder forward, it overcomes the centrifugal force on the counterweights and pulls them in toward the centerline of the propeller blades. The counterweights are attached to the blade butt so that as the counterweights move toward the blade centerline the blades are rotated to a lower angle. The propeller movement continues until the follower pins contact the stops inside the counterweights.

To increase blade angle the cockpit control is moved rearward and the selector valve is rotated to release the oil pressure from the propeller. The centrifugal force on the counterweights is now greater than the force of the oil in the propeller cylinder and the blades rotate to a higher blade angle. The oil is forced out of the cylinder and is returned to the engine sump as the cylinder is pulled inward by the action of the counterweights. The propeller movement continues until the follower pins contact the stops. The propeller is now held in high blade angle by centrifugal force acting on the counterweights.

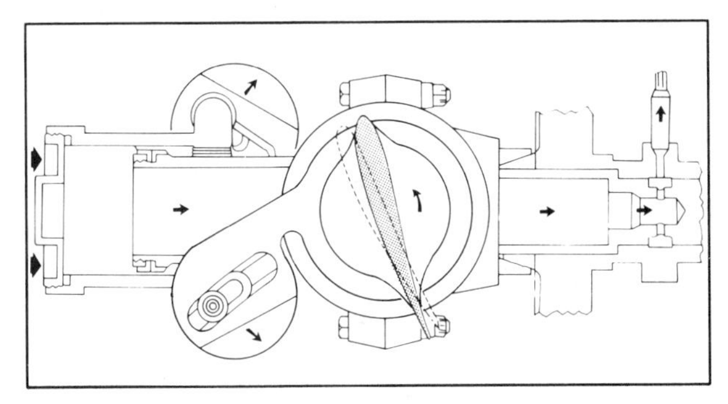

Figure 9-5. Oil pressure is released and the centrifugal force on the counterweights increases the propeller's blade angle.

C. Flight Operation

When the engine is started, the propeller is at the high blade angle setting. This is to prevent oil from going into the propeller cylinder rather than to the engine bearings and causing unnecessary bearing wear. When the engine oil temperature and pressure are at the desired values the propeller can be placed in the low blade angle position.

During engine preflight checks the propeller should be operated through at least three full pitch-change cycles to be sure that sufficient warm engine oil is in the propeller and to check for proper propeller operation.

The propeller is placed in low blade angle for takeoff and climb so maximum RPM and thrust can be developed during the low speed phases of flight. The propeller is shifted to high blade angle for cruising flight so that the maximum airspeed can be obtained.

For approach and landing, the throttle setting is reduced and the propeller is shifted to the low blade angle. This allows the full engine RPM to be available if the landing must be aborted.

When the engine is to be shut down, the propeller should be placed in high blade angle so that the majority of the oil is forced out of the propeller cylinder. This prepares the propeller for the next engine start, covers the piston surfaces with the cylinder to prevent corrosion and dirt accumulation on the piston, and prevents congealing of the oil in the cylinder when operating in cold climates.

D. Installation

1. Propeller

Hamilton-Standard propellers are used on splined shaft engines. The standard checks are made for proper front and rear cone contact and spline wear. The rear cone and propeller are placed on the shaft in the same manner as for any propeller previously discussed.

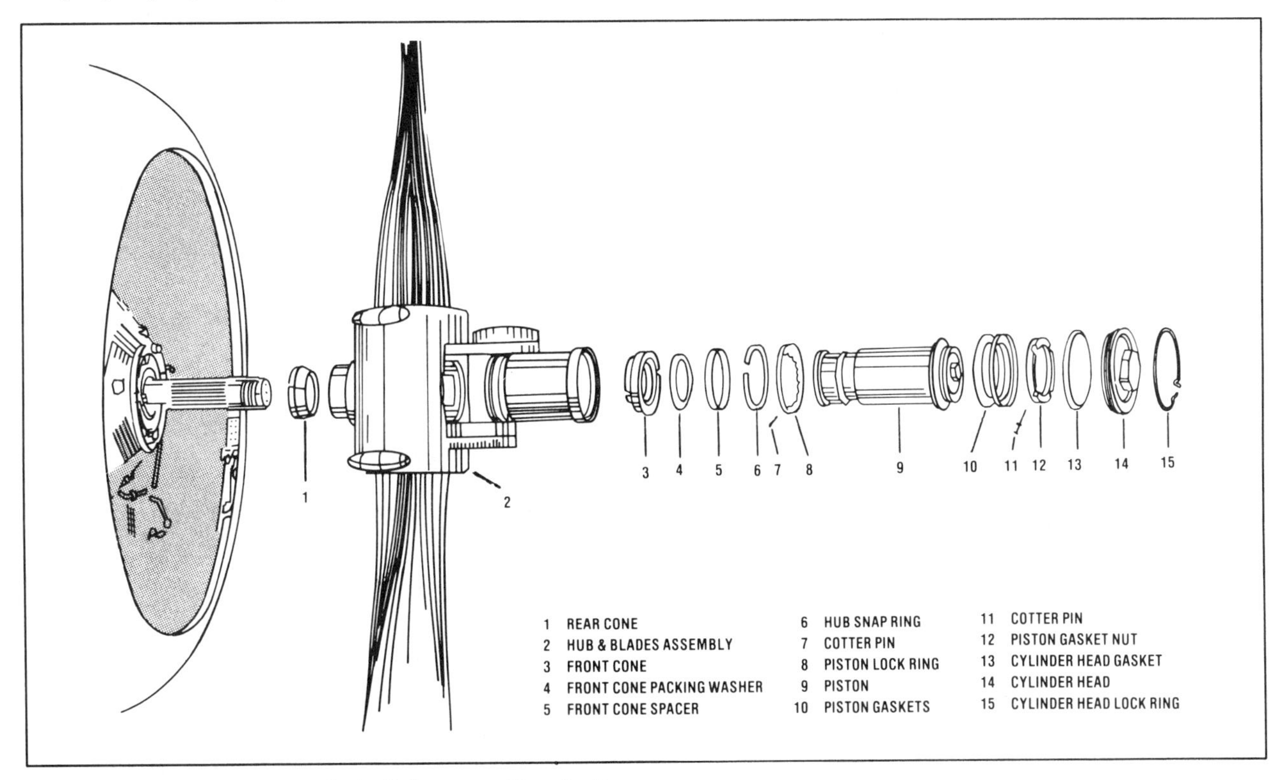

Figure 9-6. Propeller extended off the propeller shaft.

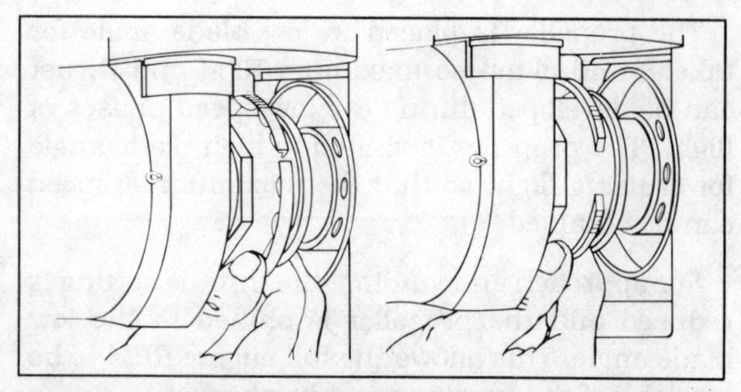

Figure 9-7. Positioning piston lock ring and hub snap ring.

The propeller piston is not installed until the propeller is placed on the shaft. The cylinder is pulled forward and the piston is inserted through the cylinder. The piston lock ring and the snap ring are then placed on the portion of the piston which was inserted through the cylinder. The front cone halves are placed on the piston and the piston is started on the threads of the crankshaft. The proper Hamilton-Standard tool is used to tighten the piston on the shaft. If the threads are damaged or the cylinder is cocked, the piston may not start on the shaft. If the piston does not turn smoothly and easily onto the shaft do not proceed until the cause of the resistance is determined and corrected. *Remember, if the threads on the crankshaft are damaged the crankshaft must be replaced!* Install seals that are required for a particular installation when mounting the propeller.

When torquing the piston refer to the propeller or aircraft maintenance manual. A specific torque wrench reading may not be given. Instead, a procedure similar to the following may be specified: apply a force of 180 pounds to the end of a four-foot bar and strike the bar once with a $2^1/_2$ pound hammer while applying the torque.

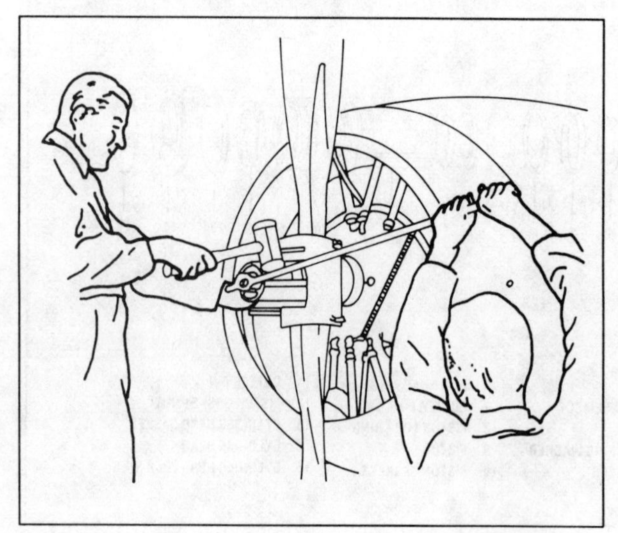

Figure 9-8. Method of torquing the propeller piston.

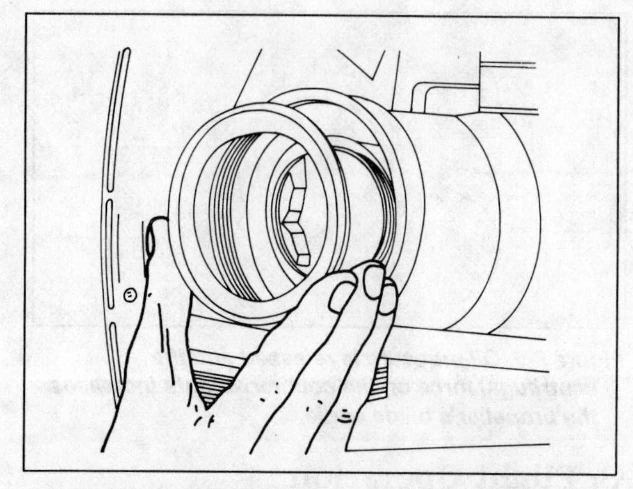

Figure 9-9. Installing a piston-to-cylinder seal.

The snap ring and piston lock ring are then installed and the lock ring is safetied to the spider with a cotter pin. The piston-to-cylinder leather seals are installed through the front of the cylinder along with the piston gasket nut. The nut is torqued and safetied.

Install the copper-asbestos cylinder head gasket with the slit toward the cylinder. Install and tighten the cylinder head and safety the cylinder head with the wire lock ring.

The propeller installation is now complete. The propeller should be checked for proper track in accordance with the aircraft or propeller maintenance manual. The propeller is now ready to have the blade angle set as discussed in Section E *Propeller Blade Angle Adjustments.*

After the blade angles have been set, the propeller should be checked for proper operation. Initially the propeller may operate erratically because of air trapped in the cylinder. This condition will correct itself as the propeller is cycled several times and the air is purged from the system.

Always refer to the propeller or aircraft maintenance manual for specific information concerning the installation of a particular model of the counterweight propeller.

2. Selector Valve And Cockpit Control

The selector valve is installed following standard procedures for the installation of engine accessories. The external oil lines should be installed in accordance with accepted aircraft practices.

The cockpit control arrangement will vary with different aircraft designs, but standard aircraft installation procedures should be followed when replacing cables, turnbuckles, pulleys, etc.

E. Propeller Blade Angle Adjustments

The propeller blade angles are adjusted by means of the stop nuts on the counterweight adjusting screw located under each counterweight cap.

The counterweight adjusting screws are removed by first removing the clevis pin which safeties the counterweight cap and unscrewing the cap. The counterweight adjusting screw is now pulled out of its recess in the counterweight or pushed from behind the counterweight bracket with a small-bladed screwdriver. The counterweight adjusting screw removal is easiest when the blades are in some mid-range position.

Alongside the recess which held the index pin is a scale which is calibrated in degrees and half-degrees with an arbitrary scale ranging from zero to ten. This scale is used to set the stop nuts on the index pin.

The propeller ***blade index number*** (also known as the base setting) should be stamped in a lead plug located near the counterweight adjusting screw recess. This number indicates the maximum blade angle for which the propeller was adjusted during its last overhaul. This number may be somewhere near 25 degrees and is used to calculate where the stop nuts on the counterweight adjusting screw should be positioned.

If the blade index is 25 degrees and the desired blade angles listed in the aircraft specifications are 17 and 22 degrees, the calculation is done in the following manner: 25 – 17 = 8 and 25 – 22 = 3.

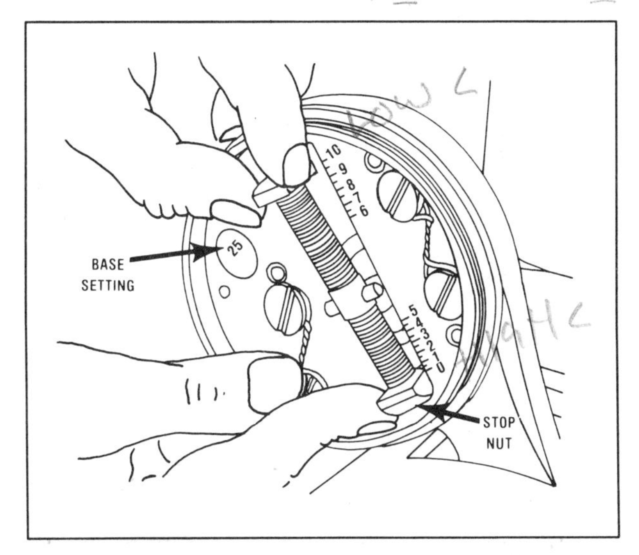

Figure 9-10. Removing the counterweight adjusting screw from the propeller counterweight.

To set the low blade angle (17 degrees) the stop nut is positioned on the counterweight adjusting screw so that the edge toward the center of the counterweight adjusting screw will line up with the 8 mark on the scale beside the counterweight adjusting screw recess. The other stop nut is positioned so that the edge lines up with the number 3. Set the stop nuts for each counterweight in this manner. With these settings the propeller blades should have the approximate blade angles desired. Due to the design of the counterweight adjusting screw and the counterweight's recess, another stop nut will have to be installed on the high blade angle end. This is to prevent the counterweight adjusting screw from cocking during operation and causing the propeller to jam. Any time that a stop nut must be positioned three degrees or more from the end of the counterweight adjusting screw, a third stop nut must be installed. This is done to all counterweights on the propeller.

Once the counterweight adjusting screws are set, the counterweight cap is screwed on. The blades are moved through their full range of travel once and then positioned for high blade angle. (Use a ***blade paddle*** on each blade when rotating the blades.) Measure the angle of each blade at the proper reference station (42 inches on most models) with a propeller protractor.

Now turn the blades to full low blade angle and measure the blade angles, which may be off as much as a full degree at this point. The angles can be corrected by placing the propeller in mid-range, carefully removing the counterweight adjusting screws so as not to disturb their adjustment, and adjusting the stop nuts. A good rule of thumb is to turn the stop nut one-quarter turn for each 0.1 degrees change in blade angle desired. Reinstall the counterweight adjusting screws and counterweight caps and measure the blade angles. Repeat as often as necessary to get the blade angles within acceptable limits.

The tolerance for blade angle adjustments are — within 0.3 degrees of the desired blade setting for the low blade angle; the blades must be within 0.2 degrees of each other at the low blade angle; the high blade angle of each blade must be within 0.1 degrees of the desired blade angle. Some installations will have different tolerances. Refer to the aircraft maintenance manual.

Once the blade angles are set correctly, reinstall the counterweight cap clevis pins and safety them with a cotter pin.

F. Inspection, Maintenance And Repair

The inspection and repair of the blades and barrel are the same as for other propeller designs. The counterweights should be checked for security and the pitch-changing mechanism should be checked for excessive play by holding one blade and noting how much the other blade can be rotated. Take care not to apply excessive force when making this check. Play between the blades may indicate wear in the blade retention area of the barrel, in the area where the follower pin moves in the counterweight, the bearing surface on the cylinder where the counterweight rides, or deterioration of the piston-to-cylinder seals. Piston-to-cylinder seals may be replaced in the field. Other defects should be referred to an overhaul facility for correction.

Since oil pressure is used to change the propeller blade angles, the areas of a leak should be noted to aid in determining the cause. If oil covers all of the propeller, from the cylinder rearward, the cylinder head is loose or the gasket is defective. If oil is found on everything aft of the cylinder, the piston-to-cylinder seals are the cause and should be replaced. Check to see that the piston is not pitted or covered with dirt. This will only damage the next set of seals. The piston should be replaced if pitted or scratched.

Oil on the barrel and blades indicates defective seals where the piston attaches to the crankshaft, a loose piston, or a crack in the spider or crankshaft. Repair or replace as appropriate.

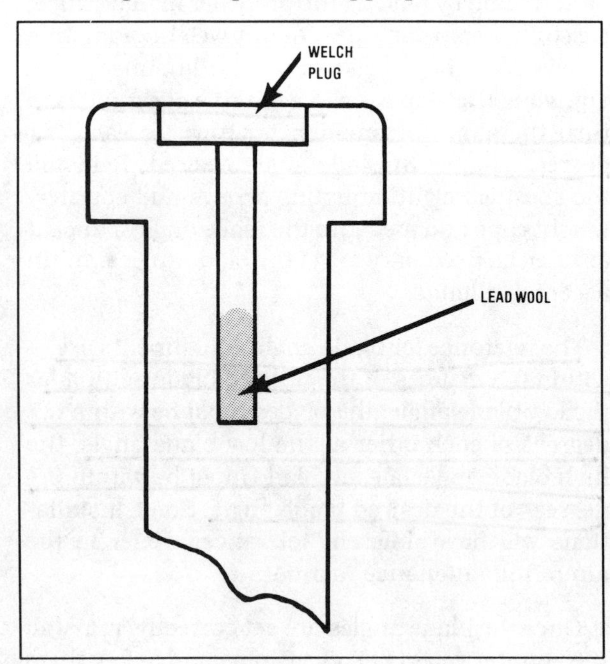

Figure 9-11. Cross section of a Welch bolt.

The counterweight propeller is balanced during overhaul by installing lead washers in the shank of a light blade for horizontal balance and placing lead washers in recesses on barrel support blocks for vertical balance.

Slight adjustments to the propeller balance are corrected by placing lead wool in the counterbored area of the hub bolt heads. These bolts are called Welch bolts because of the Welch plug used to cap the bolt once lead wool is installed. Welch plugs should be in the bolt heads regardless of whether or not lead wool is placed in the bolt head. If a Welch plug is missing, the propeller can be considered out of balance. Adjustment of balance with lead wool and Welch plugs is normally referred to an overhaul facility.

The use of lead washers and lead wool to correct propeller balance is called *dry balancing*. *Wet balancing* occurs when the blades are greased through hub fittings as the final step in overhauling the propeller. Since routine maintenance requires that the mechanic grease the blades, care should be taken to see that an equal amount of grease is used on each propeller blade. This can best be done by giving each blade the same number of strokes of the grease gun. *Do not over-lubricate the blades as this will cause the blade grease seals to fail, allowing grease to spread out over the blades and the airframe.* These seals must be replaced by an overhaul facility. The blades are greased at 25-hour intervals unless otherwise specified in the aircraft manuals.

The counterweight bearings should be lubricated in accordance with the operating conditions and the manual specifications.

G. Troubleshooting

Vibration may be investigated and corrected as has been discussed in previous chapters. The only additional cause of vibration would be unequal lubrication of the blades.

If the propeller does not respond to cockpit control movements, the following system components should be checked: the control cable to the selector valve may be rigged improperly; the selector valve may be defective and in need of repair or replacement; oil may be congealed in the oil lines or the propeller if operating in cold climates; the propeller cylinder and piston may have a build-up of sludge. The selector valve may be replaced or overhauled by a powerplant technician. The engine and propeller may be preheated to break-up congealed oil.

The propeller should be removed from the engine and all of the sludge should be removed with approved solvent (relubricate the propeller before returning to service).

If the RPM is incorrect under static or flight conditions, the blade angles may be at the wrong setting or sludge may be building up in the propeller.

QUESTIONS:

1. *What size shaft does a Hamilton-Standard 2D30-145 propeller fit?*
2. *What force is used to increase propeller blade angle?*
3. *Is the cockpit propeller control moved forward or rearward to increase blade angle?*
4. *List the three reasons that a Hamilton-Standard counterweight propeller is shut down in high blade angle.*
5. *Which component of the propeller serves as the retaining nut?*
6. *What are the stop pin settings for a ten-degree range two-position propeller indexed for 28 degrees and requiring angles of 27 degrees and 21 degrees?*
7. *What is the tolerance for the high blade angle setting of the counterweight propeller?*
8. *What are Welch bolts used for?*
9. *What is wet balancing?*
10. *When must a third stop nut be installed on the index pin?*

Chapter X
Constant-Speed Propeller Systems

A constant-speed propeller system is a system in which the propeller blade angle is varied by the action of a governor to maintain a constant system RPM. The action of the governor allows the RPM to be held constant with changes in engine throttle setting and aircraft speeds.

Constant-speed systems are used on most modern medium and high performance single-engine aircraft.

A. Theory Of Operation

1. Propeller

Constant-speed propellers use a *fixed force* to cause a decrease or increase in blade angle. This force may be centrifugal force acting on counterweights, a spring, or centrifugal twisting moment on blades. The force is termed a fixed force because it is always present during operation and must be overcome to cause a change in blade angle.

The most commonly used variable force, which will cause a change in blade angle opposite to the fixed force, is oil pressure. The oil pressure is varied by the governor as necessary to adjust the blades to the desired angle.

The operational action of the propeller is similar to that of the two-position propeller, except that the blades may be at any angle between the blade angle stops.

2. Governor

The propeller governor is an RPM sensing device which responds to a change in system RPM by directing oil pressure to or releasing oil pressure from the propeller to change the blade angle and return the system RPM to the original value. The governor is set for a specific RPM by the cockpit propeller control.

The basic governor configuration contains a driveshaft which is connected to the engine drive train. The driveshaft rotates at a fixed proportional speed to the RPM of the engine (governor speed ranges from 80% to 110% of crankshaft RPM, depending on the engine model). An oil pump drive gear is located on the driveshaft and meshes with an oil pump idler gear. These gears take engine oil at engine oil pressure and boost it to the propeller operating pressure. Excess pressure built up in the booster pump is returned to the inlet side of the pump by a pressure relief valve.

The boosted oil is routed through passages in the governor to a pilot valve which fits in the center of the hollow driveshaft. This pilot valve can be moved up and down in the driveshaft and directs oil through ports in the driveshaft to or from the propeller to vary the blade angle maintaining the desired RPM.

The position of the pilot valve in the driveshaft is determined by the action of the flyweights attached to the end of the driveshaft. The flyweights are designed to tilt outward when RPM increases and inward when RPM decreases. When the flyweights are tilted outward, the pilot valve is raised. When they tilt inward, the pilot valve is lowered. This movement of the pilot valve in response to changes in RPM will direct oil flow to adjust the blade angle to maintain the selected RPM.

The movement of the flyweights is opposed by a speeder spring which is located above the flyweights and is adjusted by the pilot through a control cable, pulley, and speeder rack. When a higher RPM is desired, the cockpit control is moved forward and the

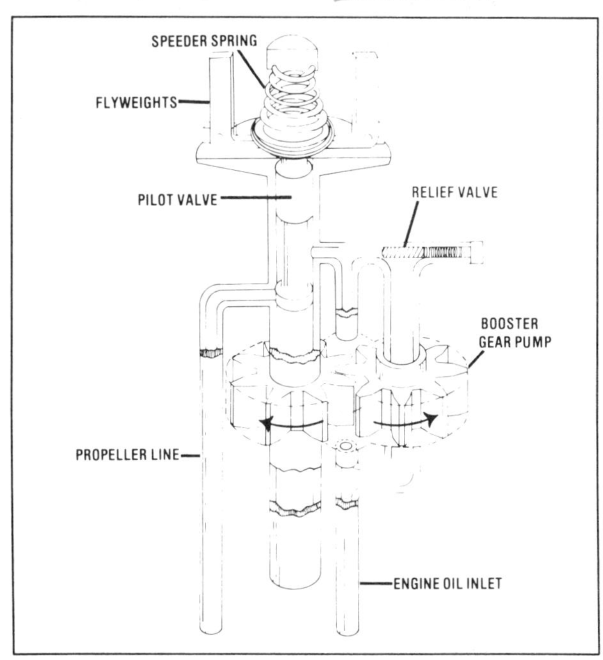

Figure 10-1. Basic governor configuration.

speeder spring is compressed. As the flyweights are tilted inward by the increase in compression of the spring, the pilot valve is lowered. When this occurs, the blade angle is decreased and the RPM will increase until the centrifugal force on the flyweights overcomes the force of the speeder spring and returns the pilot valve to the neutral position.

The opposite action will occur if the cockpit control is moved aft, the speeder spring compression will be reduced, the flyweights will tilt outward, the pilot valve is raised, and the blade angle will increase until the centrifugal force on the flyweights decreases and the pilot valve returns to the neutral position.

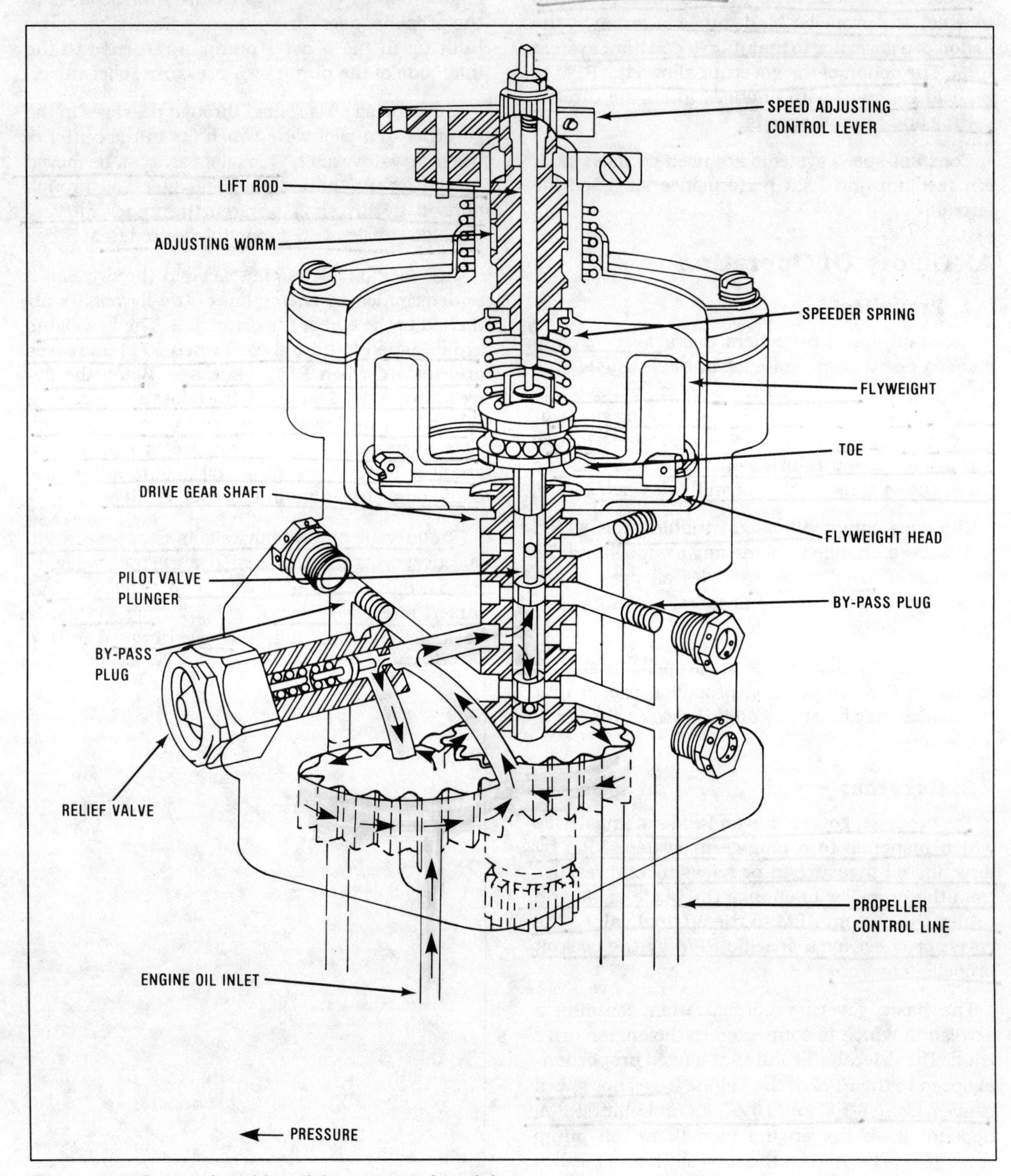

Figure 10-2. Onspeed position of the governor flyweights.

Whenever the flyweights are tilted outward and the pilot valve is raised, the governor is said to be in an *Overspeed* condition (the RPM is higher than the governor speeder spring setting). If the flyweights are tilted inward, the governor is *Underspeed* (the RPM is lower than the speeder spring setting). If the RPM is the same as the governor setting, the governor is *Onspeed.*

The same governing action of the flyweights and pilot valve will occur with changing flight conditions. If the aircraft is in a cruise condition and the pilot starts a climb, airspeed will decrease causing an increase in propeller blade angle of attack. With the increase, more drag is created and the system RPM slows down. The governor senses this reduction in RPM by the reduced centrifugal force on the

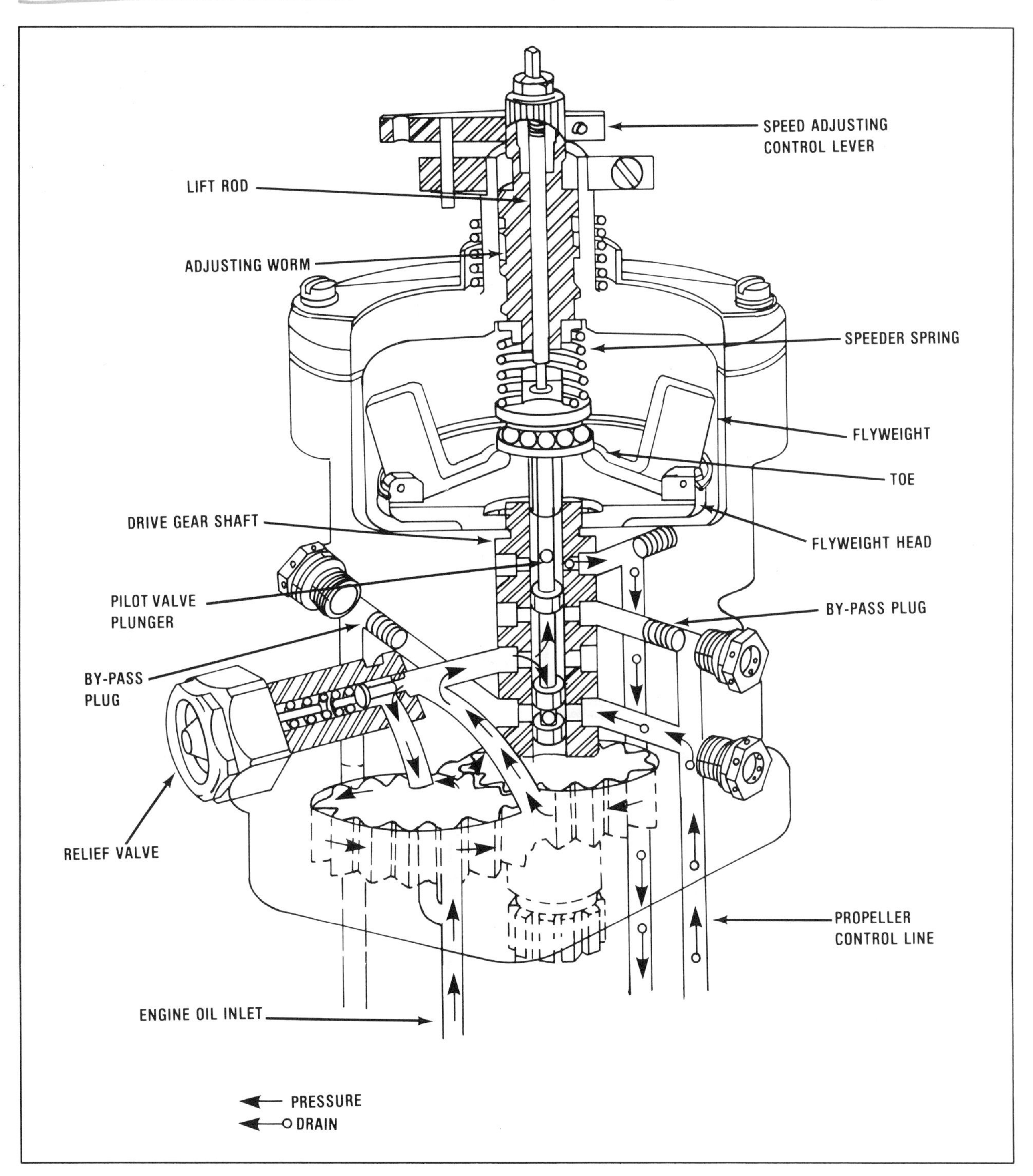

Figure 10-3. Overspeed position of the governor flyweights.

flyweights, allowing them to tilt inward and lower the pilot valve (underspeed condition). When the pilot valve is lowered, blade angle is reduced and the RPM increases to the original value, and the system returns to the onspeed condition.

If the aircraft is placed in a dive from cruising flight, an overspeed condition would be created and the governor would cause an increase in blade angle to return the system to the onspeed condition (see Figures 10-2, 10-3, and 10-4).

A change in throttle setting will have the same effect as placing the aircraft in a climb or dive. An increase in throttle would cause an increase in blade angle to prevent an RPM increase. A decrease in throttle setting would result in a decrease in blade angle.

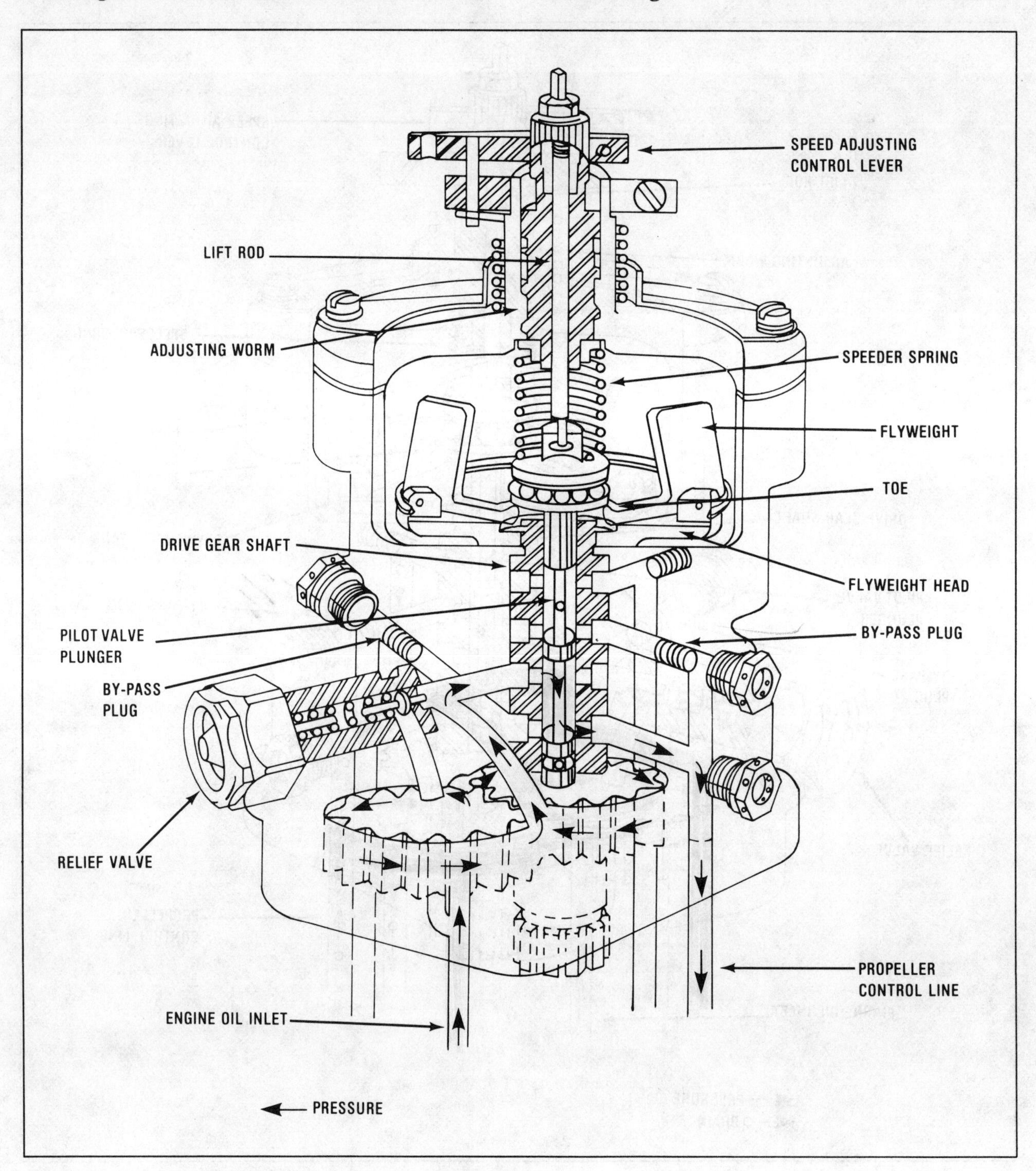

Figure 10-4. Underspeed position of the governor flyweights.

3. Instrument Indications

An aircraft with a fixed-pitch propeller uses the tachometer to indicate the throttle setting, with the RPM increasing as the throttle is advanced and decreasing as the throttle is retarded. A constant-speed system uses a manifold pressure gauge to indicate the throttle setting and the tachometer is used to indicate the setting of the propeller control. However, there is some interaction between the propeller control and the manifold pressure gauge when changing RPM and holding the throttle setting fixed.

To understand the interaction between the propeller control, tachometer, and manifold pressure gauge, it is helpful to refer to the PLANK formula used in determining Indicated Horsepower. (For a complete discussion of the PLANK formula, refer to EA-AC65-12A, *Airframe and Powerplant Mechanics Powerplant Handbook.*) The horsepower that an engine develops is a function of the engine RPM and the engine manifold pressure. The amount of horsepower (HP) produced is determined by a combination of manifold pressure (MP) and RPM.

The throttle directly controls horsepower, so with a fixed throttle setting the horsepower output of the engine is constant. For instance, assume an engine is producing 200 HP at 23 inches of manifold pressure and 2,300 RPM. If the propeller control is advanced to 2,400 RPM, the manifold pressure must decrease to some lower value, say 22 inches to agree with the formula. If the RPM is reduced to 2,200, the manifold pressure may rise to 24 inches. By these examples it can be seen that when the RPM is adjusted, a change in manifold

CONTROL SETTING	HORSEPOWER	MANIFOLD PRESSURE	RPM
THROTTLE SET AND PROPELLER CONTROL INCREASED	UNCHANGED	DECREASED	INCREASED
THROTTLE SET AND PROPELLER CONTROL DECREASED	UNCHANGED	INCREASED	DECREASED
PROPELLER CONTROL SET AND THROTTLE INCREASED	INCREASED	INCREASED	UNCHANGED
PROPELLER CONTROL SET AND THROTTLE DECREASED	DECREASED	DECREASED	UNCHANGED

Figure 10-5. A comparison of the change in manifold pressure and RPM with changes in the propeller control setting and the throttle setting.

pressure will occur. The amount of change will vary with different engines and varying situations. (The values presented here are for discussion purposes only.)

It should be noted that it is possible to damage the engine if the manifold pressure is too high for a given RPM setting. Always refer to the aircraft's engine performance charts and operating manual before operating an engine.

With a fixed propeller control setting, the manifold pressure can be changed with the throttle. The RPM, however, will remain constant because the governor will adjust the blade angle to maintain a set RPM.

When changing power settings, take care to prevent damaging the engine by creating too high a manifold pressure for a given RPM. When power is to be increased, it is best to increase the RPM to the desired setting and then advance the throttle to the desired manifold pressure. When decreasing power, pull the throttle back until the manifold pressure is about one inch below the desired setting. The propeller control can then be reduced with a rise in manifold pressure of about one inch resulting as covered in the discussion of the PLANK formula.

4. General System Configuration

The propeller cockpit control is located with the throttle and mixture controls. The control is linked to the governor by a push-pull control, torque tubes, or steel control cables. The control linkage is connected to the governor pulley or control arm and the point of attachment can be adjusted to allow the system to be properly rigged.

The governor is located under the left front cylinder on most Continental horizontally opposed engines. On a Lycoming opposed engine, the governor is commonly mounted on the engine rear accessory case. Radial engines may have the governor mounted on the nose case or on the accessory case.

The propeller installation may incorporate a spinner, crankshaft extension, and ice elimination components.

Oil passages are machined inside the engine case and carry oil to and from the propeller through the crankshaft via a transfer bearing at the engine front main journal.

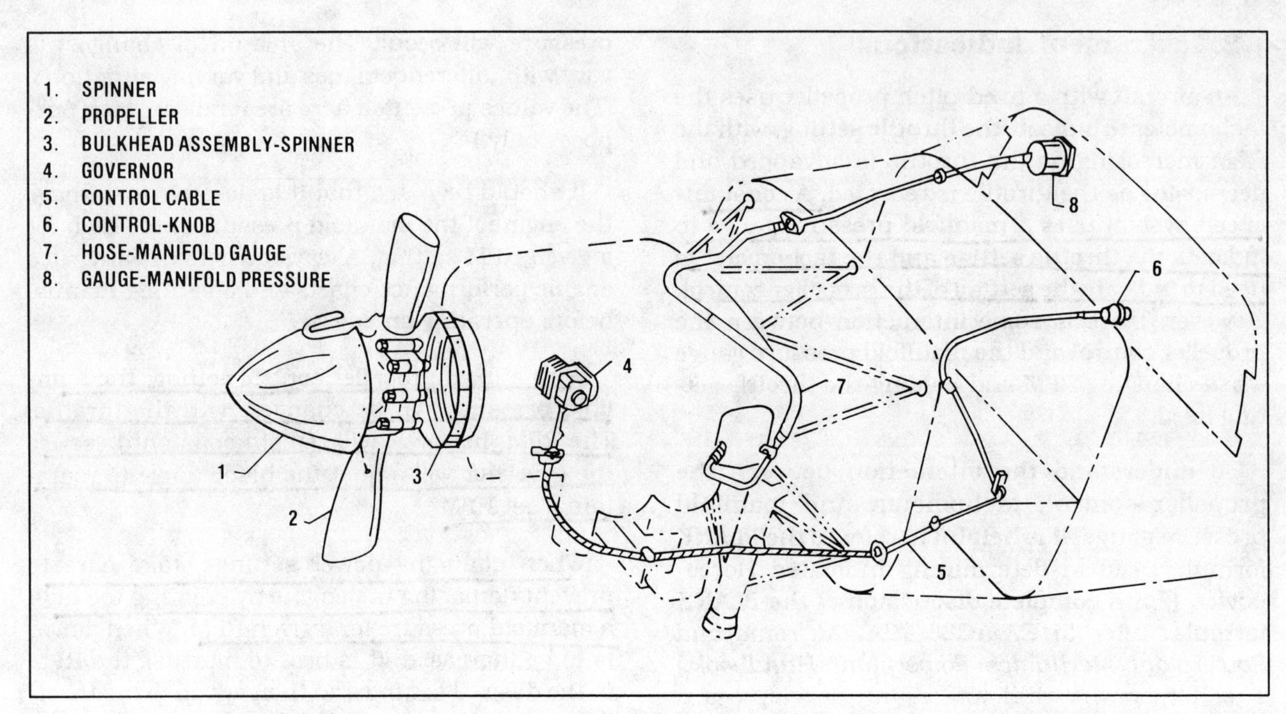

Figure 10-6. Propeller control system component arrangement for a typical light single-engine aircraft.

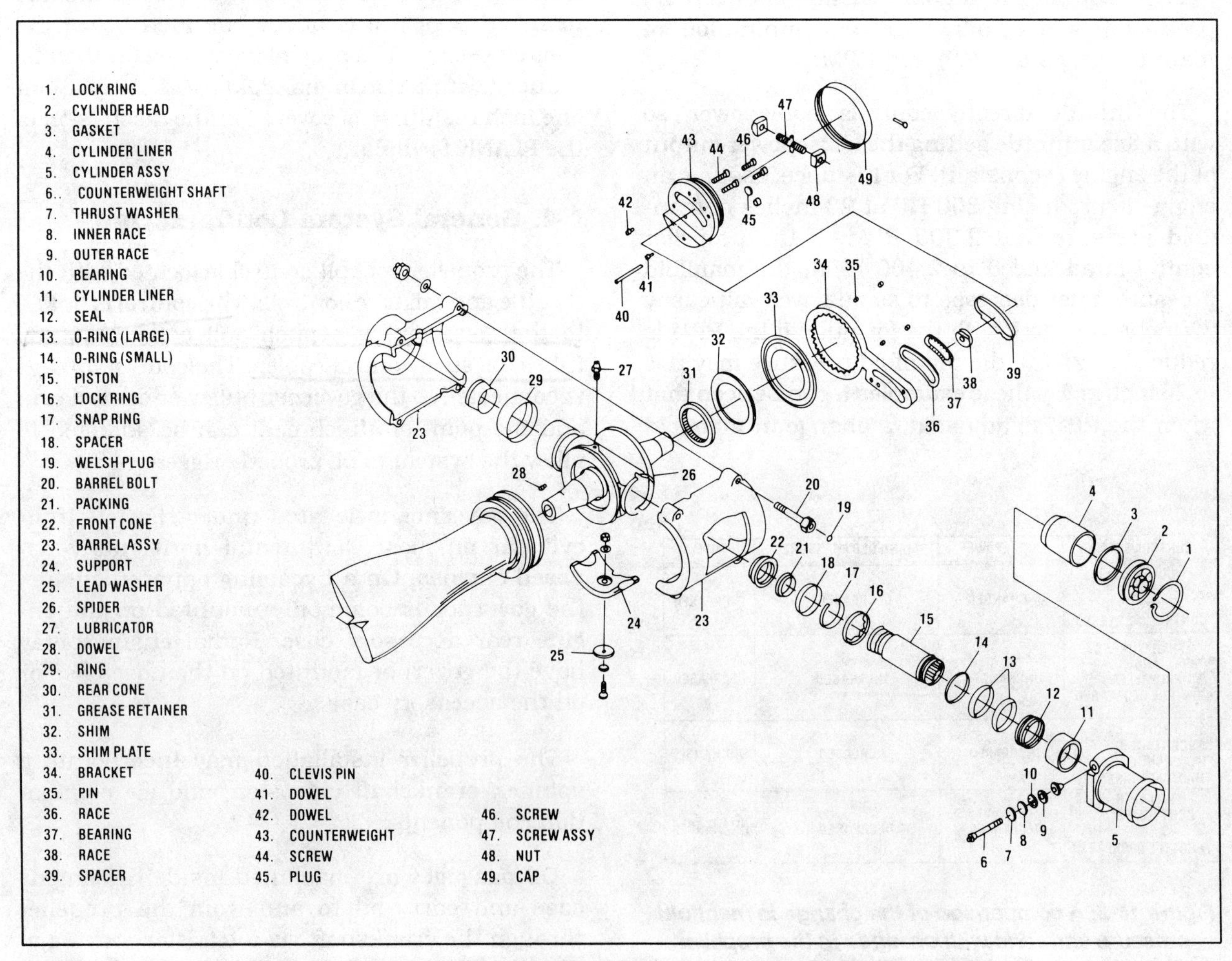

Figure 10-7. Exploded view of a Hamilton-Standard counterweight constant-speed propeller.

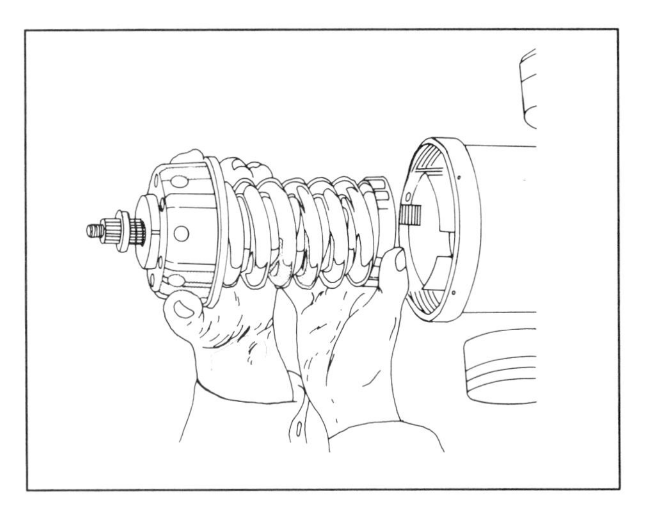

Figure 10-8. Spring assembly for a 20-degree propeller being installed.

B. Hamilton-Standard Counterweight Propeller System

The Hamilton-Standard constant-speed counterweight system is not widely used on modern general aviation aircraft except for agricultural aircraft. This system is used as the introductory system for constant-speed systems because the propeller is basically the same as the two-position propeller.

1. Propeller

The propeller is the same design as is used for the two-position system except that the constant-speed propeller has a different blade angle range. A refinement of the propeller design has a 20-degree range and uses a set of two coaxial springs mounted in the center of the piston to aid the propeller movement toward a high blade angle. The counterweight adjustment scales are stamped to reflect the 15- or 20-degree range.

The designation system for the propeller is the same as for the counterweight two-position propeller.

2. Governor

The governor is a Hamilton-Standard design and operates in a manner similar to the operation previously discussed. The governor is divided into three parts and this is reflected in the governor designation system.

The head of the governor contains the flyweights and flyweight cup, the speeder spring, a speeder rack and pinion mechanics, and a control pulley. Cast on the side of the head is a flange for the pulley adjustment stop screw. Some head designs incorporate a balance spring above the speeder rack to set the governor to cruise RPM if the control cable breaks.

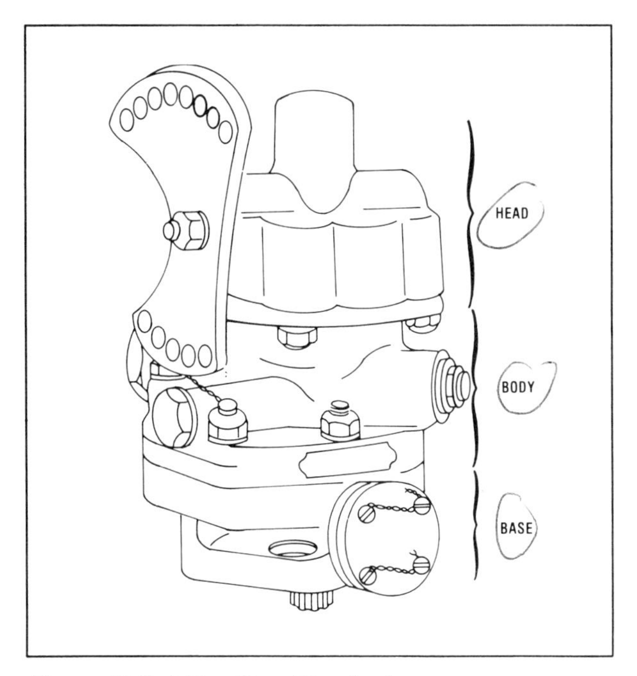

Figure 10-9. A Hamilton-Standard governor.

The body of the governor contains the propeller oil flow control mechanism, which is composed of the pilot valve, oil passages, and the pressure relief valve which is set for 180 to 200 psi.

The base contains the governor boost pump, the mounting surface for installation on the engine, and oil passages which direct engine oil to the pump and return oil from the propeller to the engine sump.

The head, body and base are held together with studs and nuts. The governor drive shaft extends below the base to mate with the engine drive gear. The driveshaft passes up through the base where it drives the oil pump, through the body where it has oil ports so oil can flow to and from the propeller, and into the head, where it is attached to the flyweight cup and rotates the flyweights.

The governor designation system indicates the design of the head, body, and base used on a particular governor. For a governor model 1A3-B2H, the basic design of the head is indicated by the *1* with minor modifications to the head design indicated by the *B* following the dash. The body design is *A* with minor modification *2*. The base is a *3* altered by an *H* minor modification.

3. System Operation

To understand the operation of the system, consider that the aircraft is in cruising flight. If the pilot desires a higher RPM from the system, he moves the propeller control forward. This rotates the pulley on the governor and causes the speeder

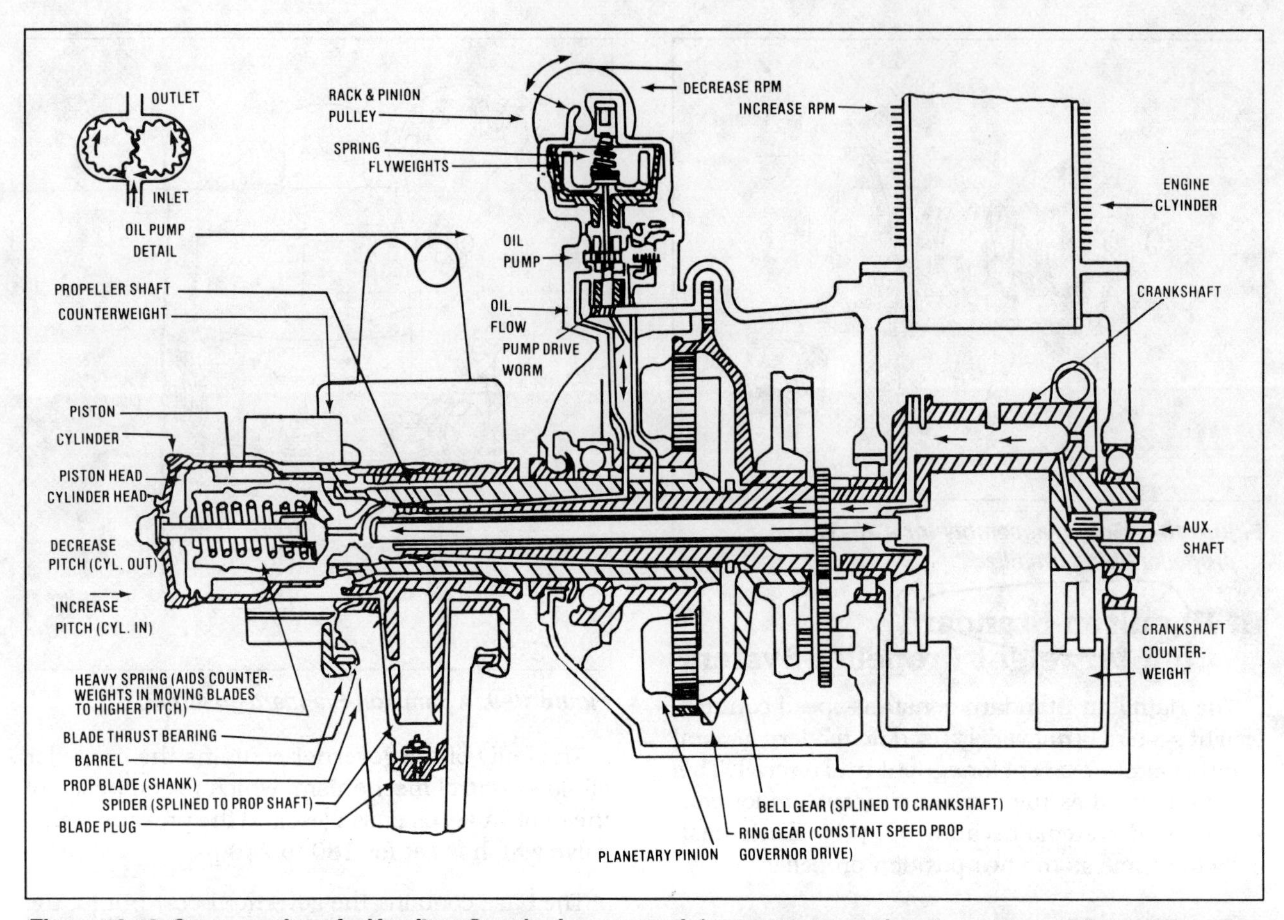

Figure 10-10. Cross section of a Hamilton-Standard counterweight constant-speed system on an engine nose case.

rack to be lowered to compress the speeder spring. As the speeder spring compresses, it presses on the flyweights and causes them to tilt inward to the underspeed position. As the flyweights tilt inward the pilot valve is lowered in the driveshaft and the port to the propeller opens, allowing oil at governor boost pump pressure to flow out to the propeller, forcing the cylinder forward and rotating the blades to a lower angle. With the lower blade angle the RPM can increase until the pilot valve returns to the neutral position and the governor is onspeed.

If the pilot decides to decrease engine RPM, he pulls the propeller control rearward. The compression of the speeder spring is reduced, allowing the flyweights to tilt outward, raising the pilot valve. The pilot valve uncovers the propeller oil passage and allows the oil in the propeller to flow into the engine sump. With no oil pressure in the propeller cylinder, the counterweights on the propeller cause the blades to increase in angle and decrease the RPM. As the RPM approaches the setting of the governor, the pilot valve returns to the neutral position and the system is again onspeed.

4. Installation And Adjustment

The propeller installation is the same as for the two-position model. The propeller is adjusted for the high and low blade angles specified for the aircraft in the same manner as for the two-position propeller. The only difference is that the calculations and settings are performed using a range appropriate for the propeller in use.

Before installing the governor, it should be inspected for freedom of rotation of the driveshaft. No binding or grittiness should be felt when rotating the shaft by hand. A smooth resistance to rotation is normally the result of the preservative compounds placed in the governor after overhaul, or is caused by cold oil in the governor, and is not necessarily an indication of a defect. The preservatives in the governor are not removed before installation unless so specified by the installation instructions. Most preservatives are compatible with the engine oil and are not in sufficient quantity to interfere with system operation.

Visually inspect the governor for surface defects, broken safeties, and security of all screws, bolts and nuts.

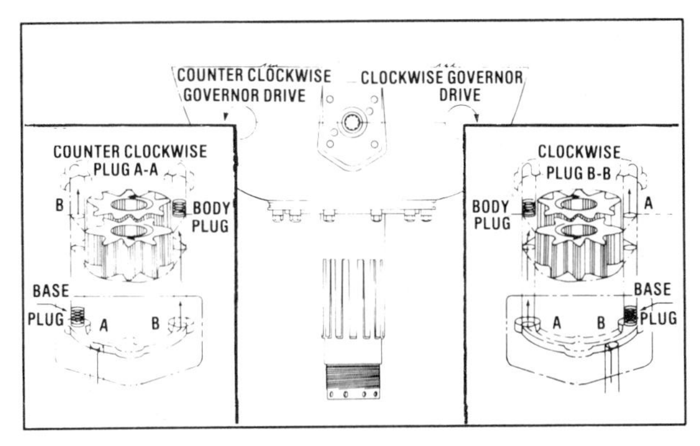

Figure 10-11. Determine if the governor oil ports are plugged properly by observing the direction of rotation of the engine drive gear.

Check to see that the correct port is plugged on the base of the governor. The governor direction of rotation can be checked by observing the direction of rotation of the governor drive gear in the engine while rotating the engine in the normal direction. Place some clean engine oil in the inlet passage on the base of the governor while holding the governor with the base up. Rotate the governor drive gear in the proper direction. If the correct port is plugged, the oil will be pulled in through the unplugged port. If the wrong port is plugged, the governor will have to be returned to an overhaul facility to have the internal ports changed.

Once the governor is inspected and the engine mounting pad is checked for scratches and burrs, the mounting gasket can be placed on the pad with the raised portion of the filter screen toward the governor. Gasket compound is not used.

(i.e. up)

Place the governor on the engine, rotating the crankshaft as necessary to allow the splines on the governor driveshaft to engage the engine gear. Install the mounting nuts and washers, torque, and safety the installation.

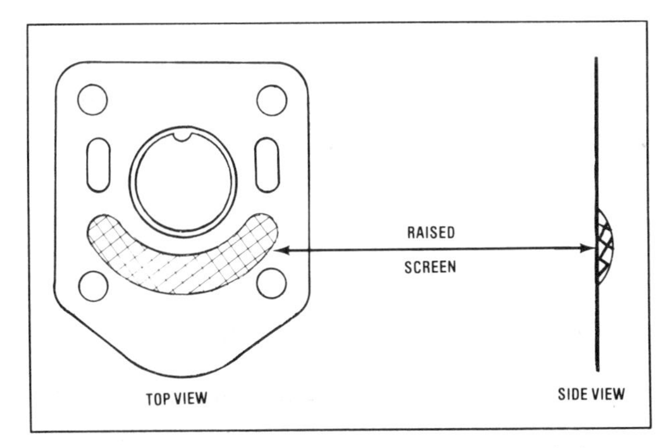

Figure 10-12. The governor mounting gasket is installed with the raised screen toward the governor.

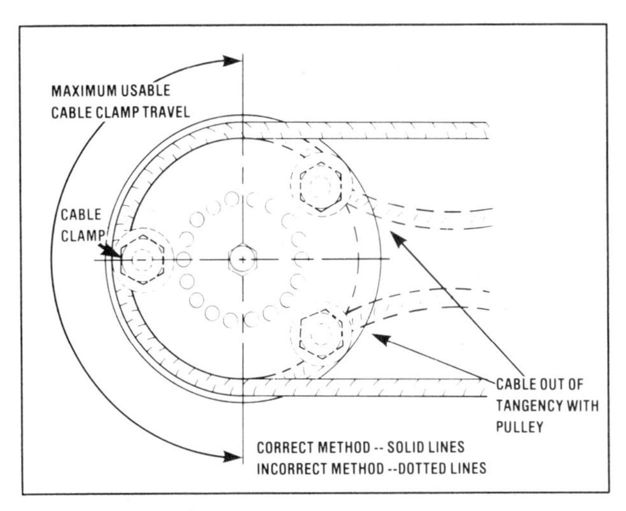

Figure 10-13. The governor pulley is installed so that the cable clamp will allow the cable to be tangent to the pulley at all times.

The governor must now be rigged to allow the proper cockpit control movement and set the desired RPM limit. Place the pulley on the hexagon shaped pinion shaft on the governor head in such a position that the cable clamp bolt hole will be opposite the direction that the control cable comes from when the governor pulley is in mid-range. This will assure that the cable remains tangent to the pulley during operation. Then install the washer, nut, and cotter pin to hold the pulley on the shaft (Figure 10-13).

Install the governor pulley stop pin in a pulley hole which will allow the pin to contact the stop screw when the governor is in the full high RPM position. The control cable is now placed around the pulley.

Move the cockpit control full forward and then back off about 1/8 of an inch to allow the proper control cushion. Lock the control in this position. Rotate the pulley to the high RPM setting and install

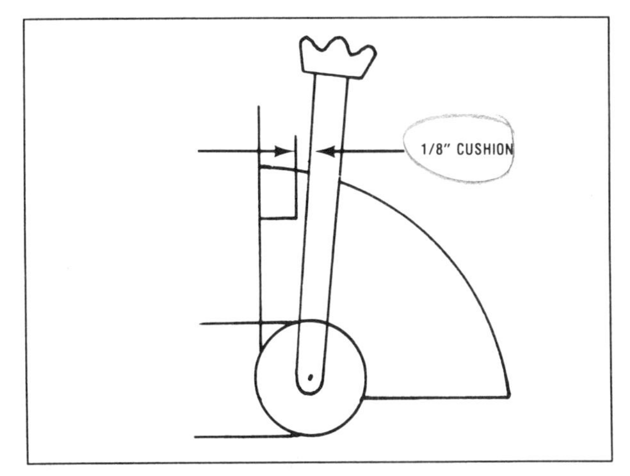

Figure 10-14. Adjust the cockpit control for a cushion of about 1/8-inch unless otherwise specified.

the cable clamp so that the pulley will move with the cable. Adjust the cable tension to the value specified for the aircraft and recheck so that the pulley stop pin touches the stop screw when the cockpit control is full forward (allowing for the cushion).

Once everything is secured and safetied as necessary for a ground operational check, check the adjustment of the governor high RPM stop by starting the engine and slowly advancing the throttle until the RPM no longer increases. (Be sure the propeller control in the cockpit is full forward). Note the RPM and stop the engine. Determine the difference between the RPM obtained and the desired maximum RPM. Adjust the governor to give the desired maximum RPM limit by moving the stop pin on the pulley one pin hole for each 250 RPM change desired. Turn the stop screw one turn for each 25 RPM change desired. Restart the engine and check the RPM. Make final adjustments with partial turns of the stop screw.

When making the initial RPM check, if the RPM continues to increase when the engine RPM limit has been reached, retard the propeller control until it starts to decrease. Shut the engine down and adjust the pulley stop pin position and the governor stop screw so that the stop pin just contacts the stop screw. Readjust the control cable for the correct control cushion. Restart the engine and adjust the governor as necessary to achieve the maximum governed RPM desired. Do not operate the engine above its rated RPM!

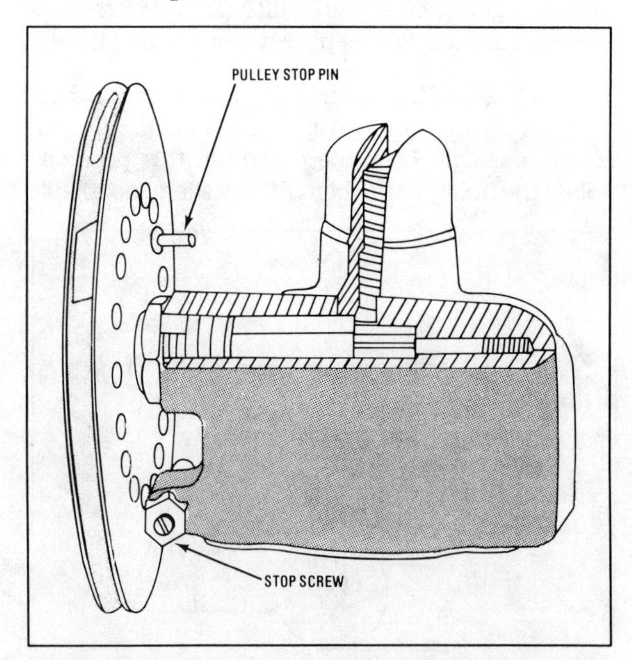

Figure 10-15. Adjust system RPM limit by moving the pulley stop pin and adjusting the governor stop screw.

All components are now checked for torque, safety, and movement in accordance with the appropriate manuals. A ground and flight check should be performed to check operation and oil leaks.

5. Inspection, Maintenance And Repair

The propeller should be treated in the same manner as discussed in the section on two-position propellers.

The governor should be visually inspected for leaks, surface defects, defective safeties, and loose nuts and bolts. If leaks cannot be corrected by properly torquing the nuts and bolts, the governor should be referred to an overhaul facility for correction.

Inspect and adjust control rigging, cushion, and cable tension as necessary, according to the aircraft maintenance manual.

6. Troubleshooting

Propeller troubleshooting is the same as for the two-position propeller.

If the propeller shifts to high blade angle during operation and does not respond to cockpit control movements, one of three problems may have developed, the governor driveshaft or engine drive gear may have sheared, oil to the governor may be blocked, or the governor speeder spring may have broken. Clear the oil passage or have the engine and/or governor repaired as appropriate.

If the governor contains a balance spring above the speeder rack, the governor will go to a cruise RPM setting if the cable from the cockpit control were to break. If a balance spring is not used, the governor pulley will be spring loaded to the maximum RPM position.

Sluggish or erratic propeller operation may be an indication of sludge build-up in the governor. The sludge can be flushed from the system by removing the governor from the engine and placing it under a low-pressure flow of solvent. Rotate the governor by hand in its normal direction of rotation and pump the solvent through the governor to cut the sludge loose. When all of the sludge has been removed, indicated by clear solvent coming out of the governor, pump clean engine oil through the governor, rotating it by hand. Reinstall the governor on the engine.

Overspeeding on takeoff may be caused by incorrect propeller or governor setting, but it may also be a result of too rapid an application of throttle. Check with the pilot and perform a ground run-up check before adjusting the governor. Also, be sure that the tachometer is accurate.

If, when troubleshooting the constant-speed system, you suspect the governor, remove the governor and replace it with a known good governor. If the problem does not go away, the governor is not the cause. The same procedure can be followed with the propeller.

C. McCauley Propeller System

The McCauley constant-speed propeller system is one of the more popular constant-speed systems for light and medium size general aviation aircraft.

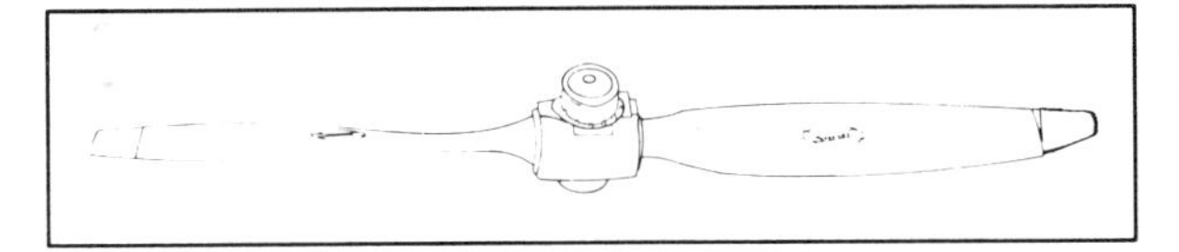

Figure 10-16. McCauley constant-speed propeller.

The system is used on most of the Cessna aircraft requiring a constant-speed system and on many Beechcraft and other designs.

1. Propellers

Two series of propellers are currently being produced by McCauley — the threaded series and the threadless series. The threaded series propellers use a retention nut which screws into the propeller hub and holds the blades in the hub. The threadless series propellers use split retainers to hold the blades in the hub. The threadless design is the more modern of the two and has the advantages of simplified manufacture and decreased overhaul time.

McCauley propellers use oil pressure on an internal piston to increase blade angle, while centrifugal twisting moments on the blades combined with a booster spring in the hub provides the pitch decreasing forces. The movement of the piston is

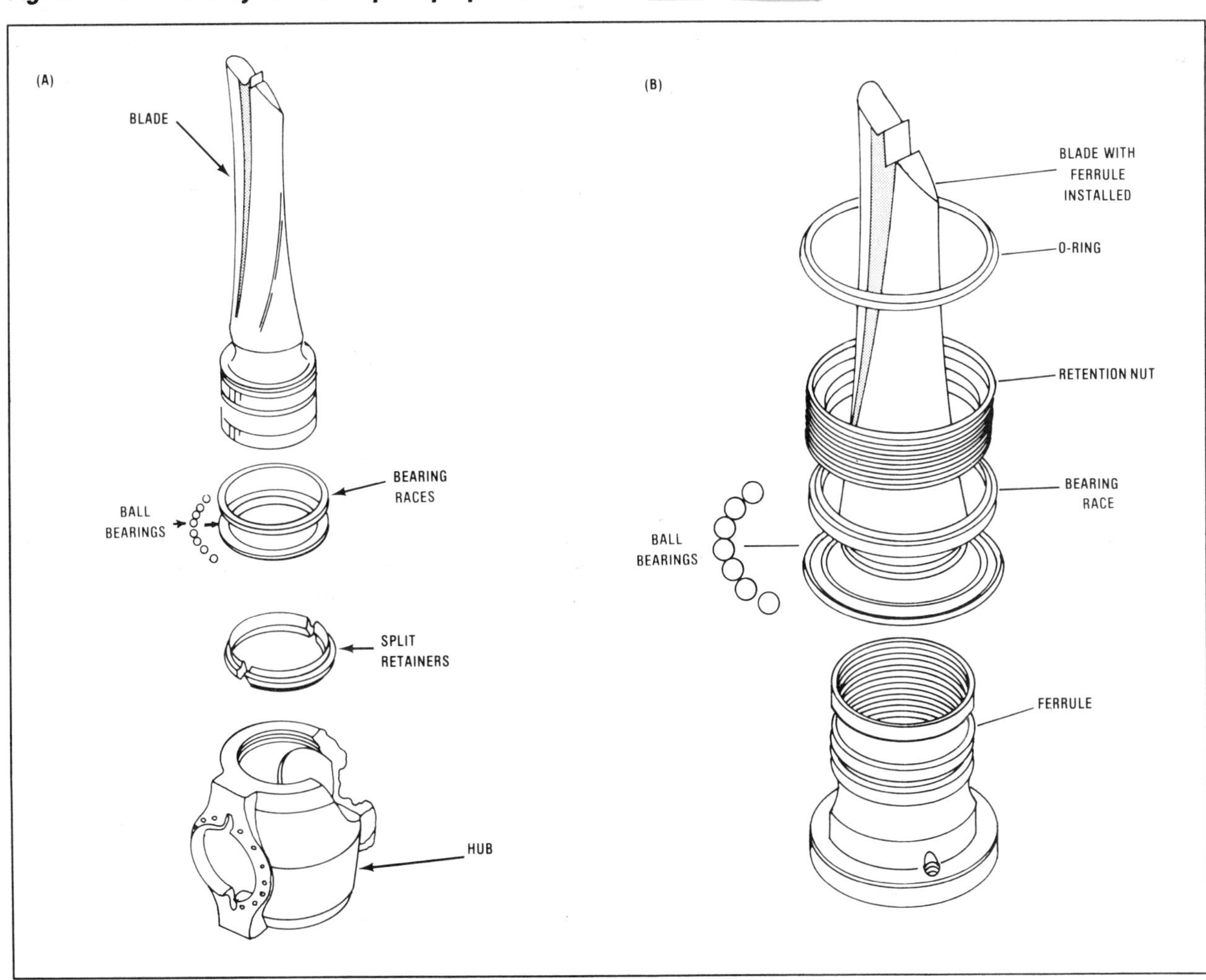

Figure 10-17A. Exploded view of a McCauley threadless blade.

Figure 10-17B. Exploded view of a McCauley threaded propeller blade.

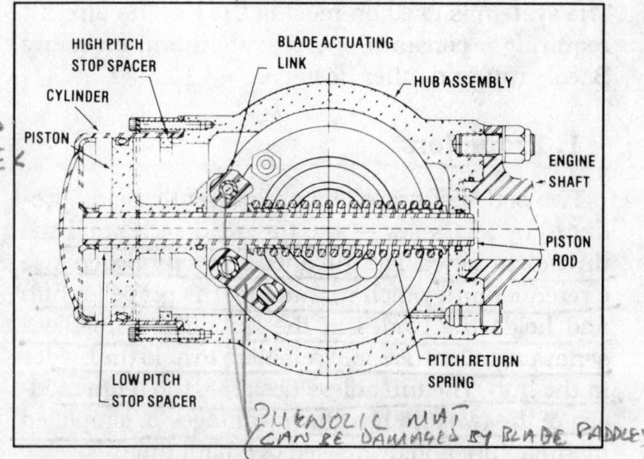

Figure 10-18. Cross section of a McCauley constant-speed propeller.

transmitted to the blade actuating pins, located on the butt of the blades, by blade actuating links. All of the pitch-changing mechanism is located inside the hub.

The propeller blades, hub, retention nut, and piston are made of aluminum. The propeller cylinder, blade actuating pins, piston rod, and spring are made of steel, plated with chrome or cadmium. The actuating links are made of a phenolic material.

The hollow piston rod through the center of the hub is used as the oil passage to direct oil from the engine crankshaft to the propeller piston. The pitch return spring is located around the piston rod and is compressed between the piston and the rear inside surface of the hub. O-ring seals are used to seal between the piston and the cylinder, the piston and the piston rod, and the piston rod and the hub.

All operating components of the propeller are lubricated at overhaul and receive no additional lubrication during operation. The hub is vented to the atmosphere through spiral pins in the hub just behind the cylinder or through holes located in the mounting dowel pins.

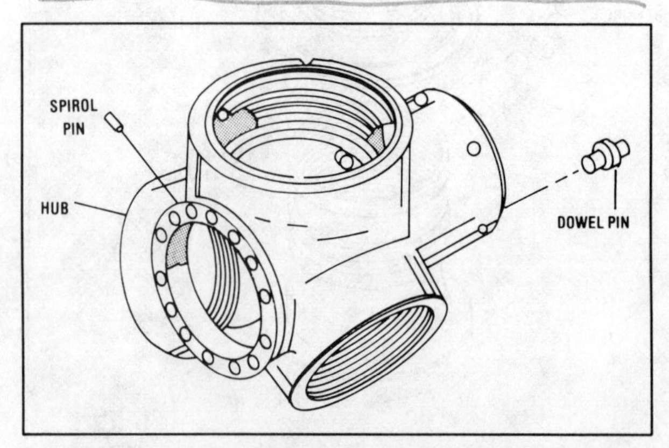

Figure 10-19. The McCauley propeller hub is vented to the atmosphere through a spirol pin on the front of the hub or a dowel pin on the rear of the hub. Only one will be used in a specific model propeller.

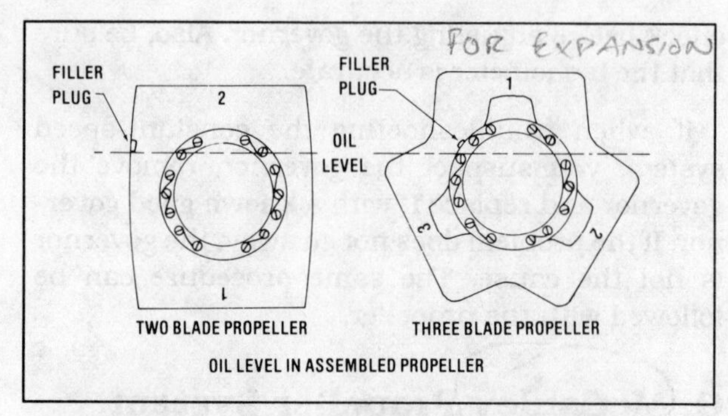

Figure 10-20. Locations of hub plugs on oil-filled propellers.

Certain models of McCauley propellers have been modified to allow for an on-going dye-penetrant type inspection. The hub breather holes are sealed and the hub is partially filled with engine oil colored with a red dye. The red dye in the oil makes the location of cracks readily apparent and indicates that the propeller should be removed from service. Cracking may occur in the blade retention area of the hub and on the blade shanks. A leaking cylinder gasket may also be indicated and can be replaced in the field. The area should be inspected carefully to determine if any cracks have developed in the cylinder mounting area.

Oil-filled hubs are readily identified by the installation of an internally wrenched pipe plug on the side of the hub.

For Continental installations, mounting studs and dowel pins are installed on the rear mounting surface of the hub. Lycoming installations use a stud and nut arrangement locked together with a roll pin so that the studs can be screwed into the threaded inserts on the Lycoming crankshaft.

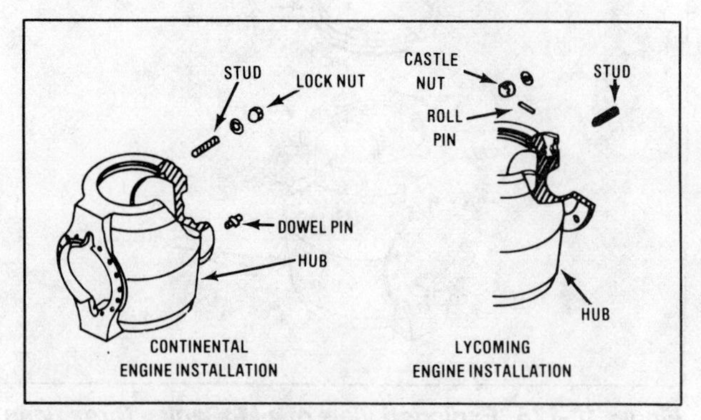

Figure 10-21. McCauley hubs designed to fit Continental or Lycoming engines.

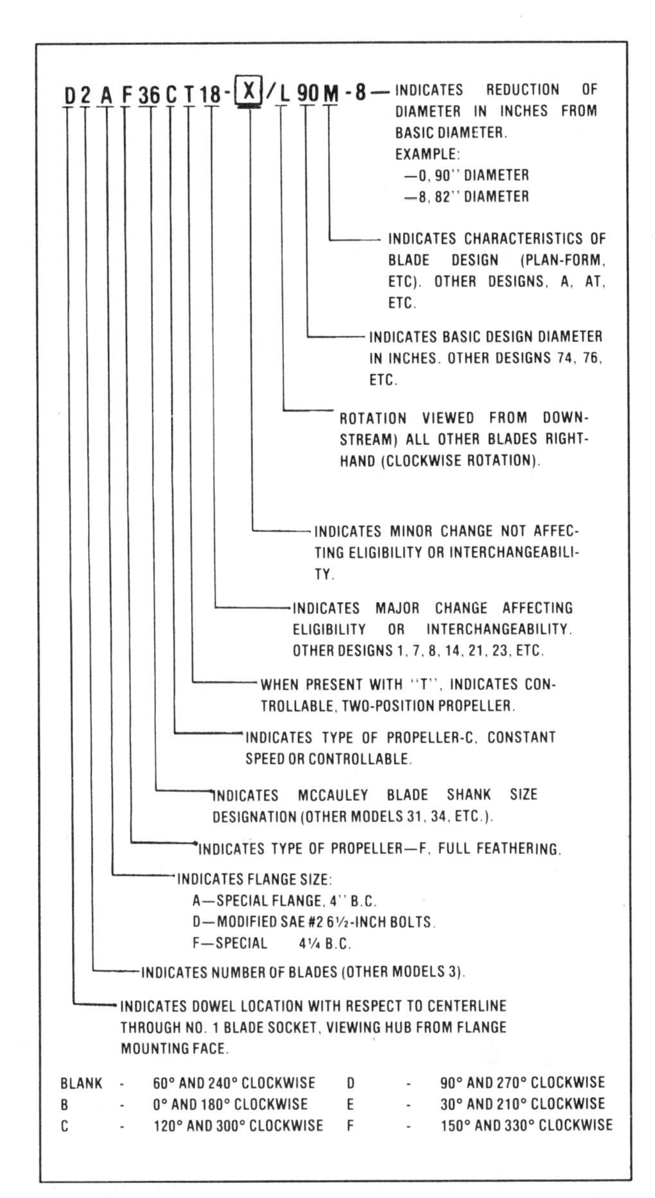

Figure 10-22. McCauley propeller designation system.

The McCauley propeller designation system is broken down in Figure 10-22. The most important parts of the designation for the technician are the dowel pin location, the C-number (C18, C22, C201, etc.), and the modification or change letter after the C-number. The modification and change designation indicates compliance with required or recommended alterations.

The blade designation is included with the propeller designation when determining which propeller will fit a specific aircraft. For example: a C203 propeller will fit a Cessna 180J aircraft, but the land-plane version requires a 90DCA-8 blade of 82 inches in diameter, while the seaplane version requires a 90DCA-2 blade of 88 inches in diameter.

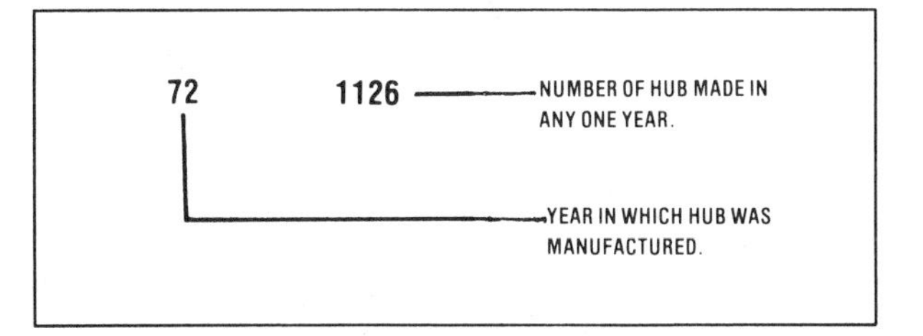

Figure 10-23. McCauley propeller serial number designation system.

McCauley serial numbers on the propeller hub indicate the year in which the hub was manufactured (Figure 10-23). If an aircraft is manufactured in 1965 shows no record of propeller replacement or overhaul, and the hub serial number is 701362, something is amiss because the hub was not manufactured until 1970!

2. Governors

McCauley governors use the same principles of operation as the Hamilton-Standard governors except that oil is released from the propeller to decrease blade angle, directly opposite from the oil flow in the Hamilton-Standard system. The governor relief valve is set for an oil pressure of 290 psi. The governor control lever is spring loaded to the high RPM setting. The overall construction of the governor is simpler than the

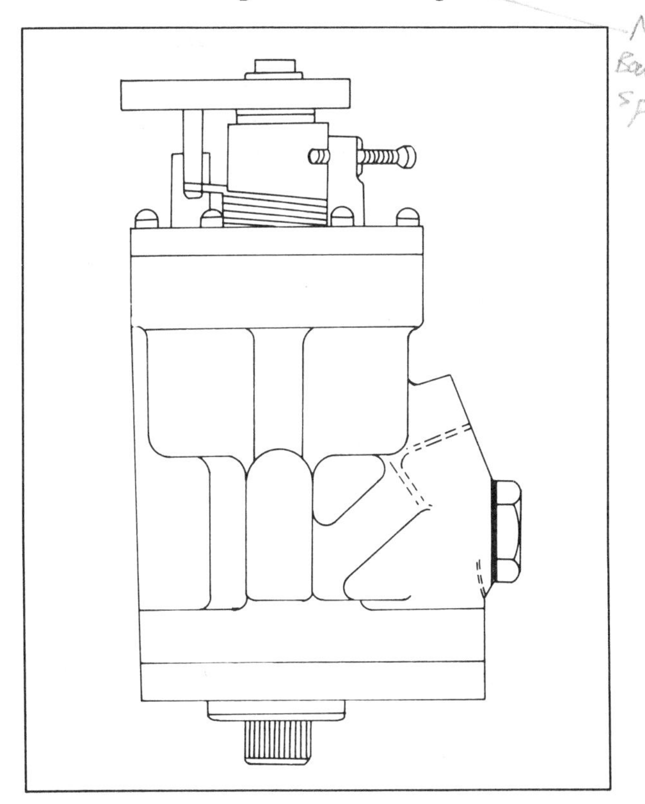

Figure 10-24. A McCauley constant-speed governor.

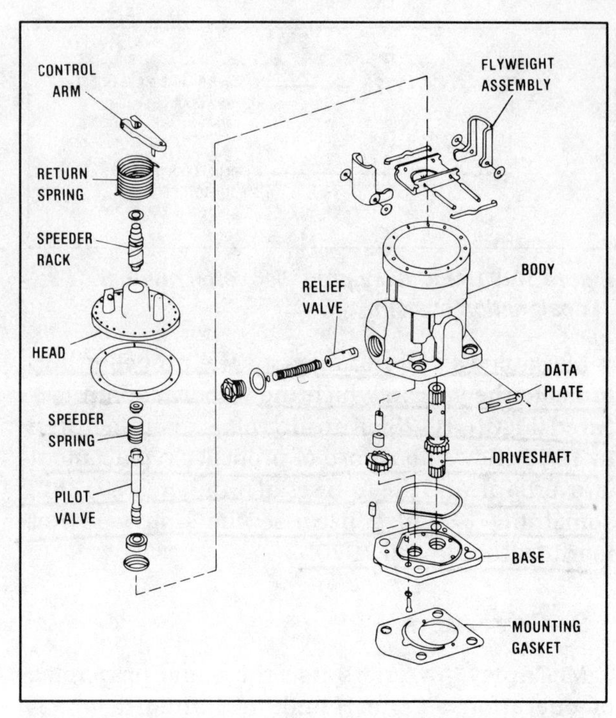

Figure 10-25. Exploded view of a McCauley governor.

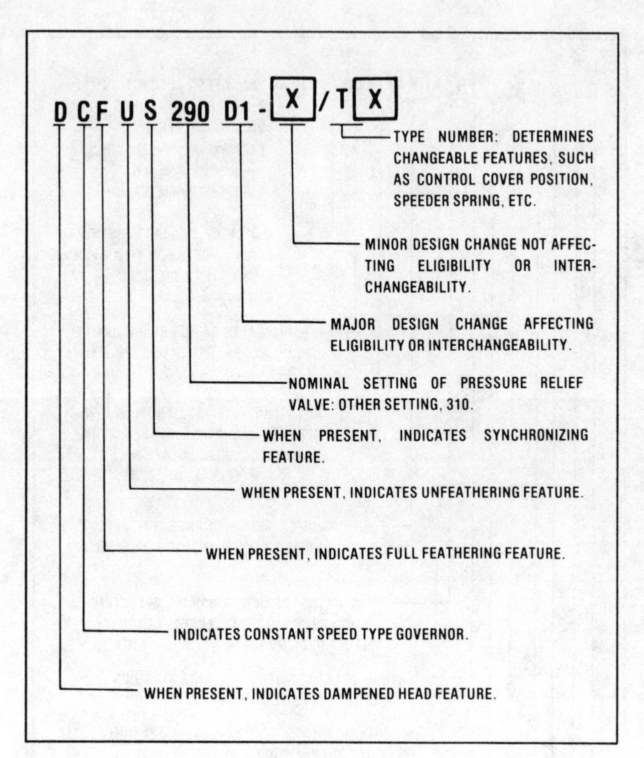

Figure 10-27. McCauley governor designation system.

Hamilton-Standard governor, being lighter and smaller. All governors incorporate a high RPM stop and some governors also use a low RPM stop. The governor designation system is illustrated in Figure 10-27.

3. Installation And Adjustment

McCauley constant-speed propellers are found only on flanged installations and are installed following the basic procedures for fixed-pitch flanked installations. Before the propeller is placed on the crankshaft, the O-ring in the rear of the hub should be lubricated with a light coat of engine oil. When placing the propeller on the flange, take care to protect this O-ring from being damaged. The bolts or nuts should be torqued following a torquing sequence and safetied as appropriate. Be sure to use new fiber lock nuts, if applicable.

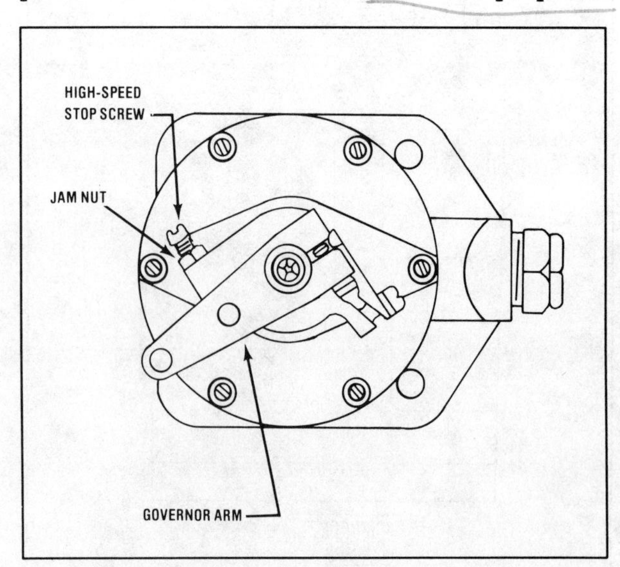

Figure 10-26. The McCauley governor high RPM stop screw is located on the head of the governor.

The governor is installed using the procedure covered in the Hamilton-Standard constant-speed section. The McCauley governor uses a control arm instead of a pulley to connect the governor control shaft to the cockpit control cable. The push-pull cockpit control is adjustable in length through a limited range by an adjustable rod end. The governor RPM limit can be adjusted by the set screw on the head of the governor with one turn of the screw changing the RPM by 17, 20, or 25 RPM, depending on the engine gear ratio and the governor.

Once the propeller and governor are installed, the cockpit control is rigged for the proper cushion. Check the system for proper operation on the ground, correcting adjustments and rigging as necessary for maximum RPM limit, response to cockpit control movement, oil leaks and control cushion. A test flight should now be performed to check operation and to check for oil leaks.

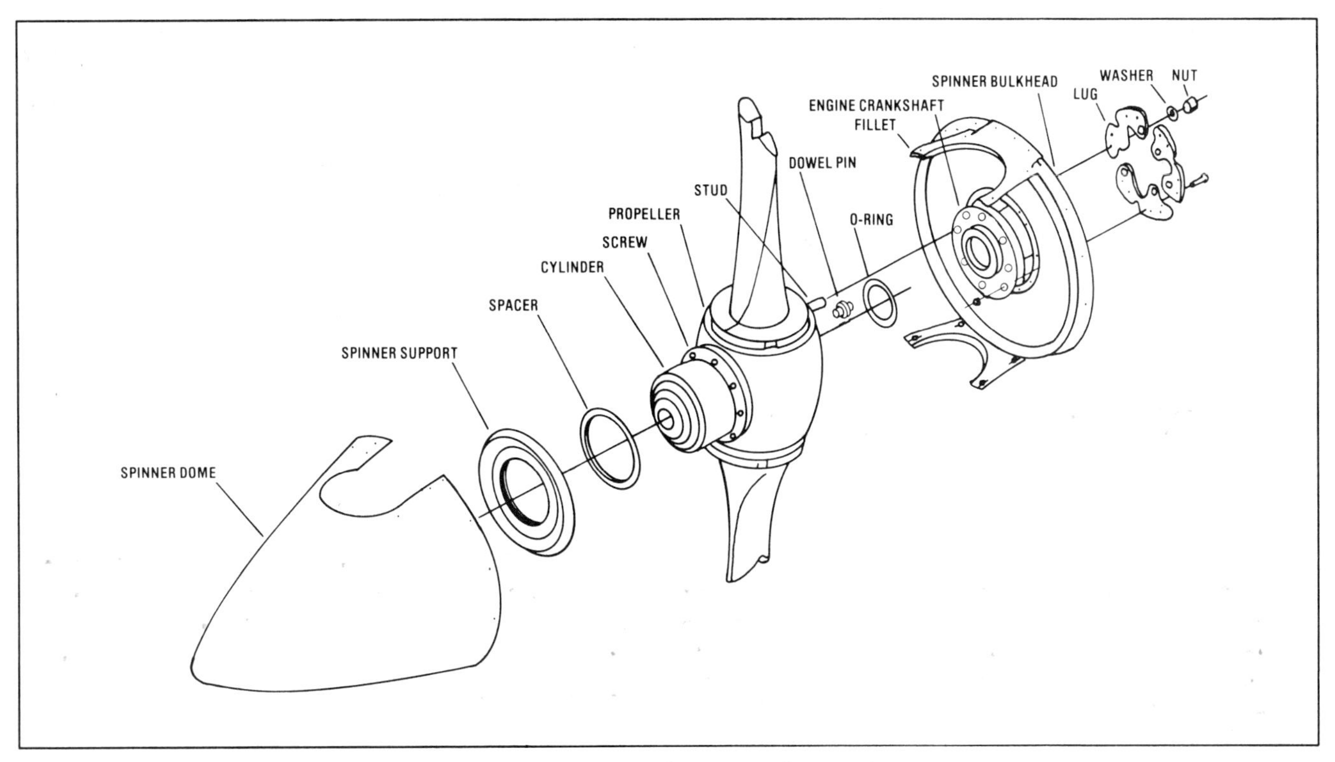

Figure 10-28. Exploded view of a McCauley constant-speed propeller installation.

After the test flight, check all nuts and bolts for torque and safety. Make sure that oil is not leaking from between the propeller hub and crankshaft. If oil is found in this area, the O-ring that seals between the crankshaft and the hub may be damaged and will need to be replaced. Do not release the aircraft for flight until the aircraft is test flown to check operation and oil leaks. The oil leakage can quickly cover the windscreen and the pilot will not be able to see out the front windscreen. For this reason, keep the test flight near the airport!

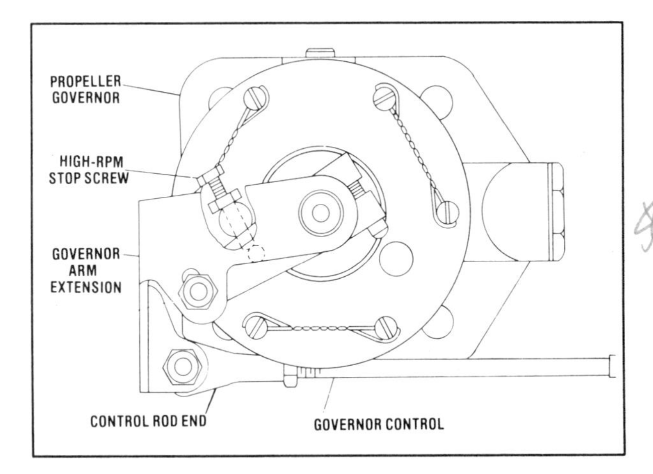

Figure 10-29. The governor control rod end position on the governor control rod can be varied by screwing the rod end on or off the threaded portion of the governor control.

4. Inspection, Maintenance And Repair

The propeller should be inspected for surface defects on the blades and hub areas, security of the blades in the hub, proper safety installation, oil leaks, and security of mounting bolts and nuts. Repair blade defects following the procedures discussed in Chapter IV. Repair defects on the surface of the hub using procedures in the McCauley maintenance manuals with special note taken of the location and size of the defect. (Certain areas of the hub are critical and are not repairable.) The track and play in the blades may be corrected by an overhaul facility if beyond the limits set by the manufacturer for each model propeller.

Oil found coming from the hub breather holes (spiral pin or dowel pins) indicates a defective piston-to-cylinder O-ring. On some models, this can be replaced by the mechanic following the propeller or aircraft maintenance manual. Otherwise, refer the propeller to a repair facility.

A dye-penetrant inspection of the blade retention areas of the hub and of the blade shanks is advisable at each 100-hour and annual inspection. Be sure to remove all of the residue after the inspection as some of the substances used are corrosive. Cracked components should be referred to a repair station for replacement.

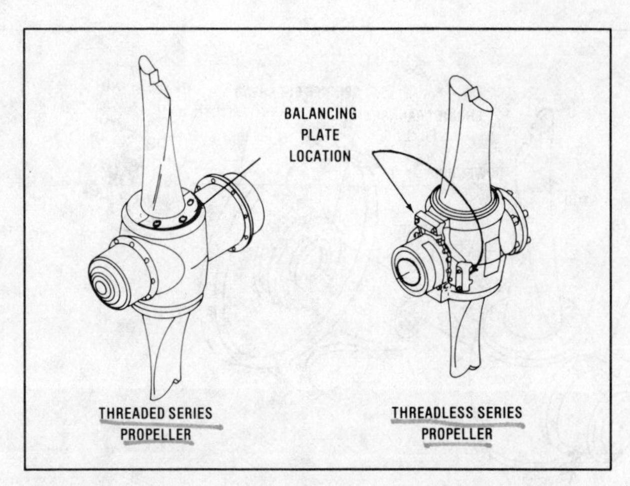

Figure 10-30. Location of balance plates on McCauley constant-speed propellers.

The McCauley threaded series propellers are balanced by the installation of balance plates on the blade retention nuts. Threadless propellers are balanced by the use of balance plates around the cylinder. Inspect the plate installations for security and safety (Figure 10-30).

Propeller designs which use a spiral pin should have a coat of General Electric RTV-108 silicone base sealer placed around the outside of the pin to prevent water from entering the hub.

Typically McCauley propellers are overhauled at 1,200-hour intervals or whenever the engine is overhauled, whichever comes first. The governors are overhauled at 800-hour intervals or engine overhaul, whichever comes first. Maintenance manuals should be consulted for correct intervals.

5. Troubleshooting

The troubleshooting procedures for the McCauley constant-speed components are basically the same as for the Hamilton-Standard system.

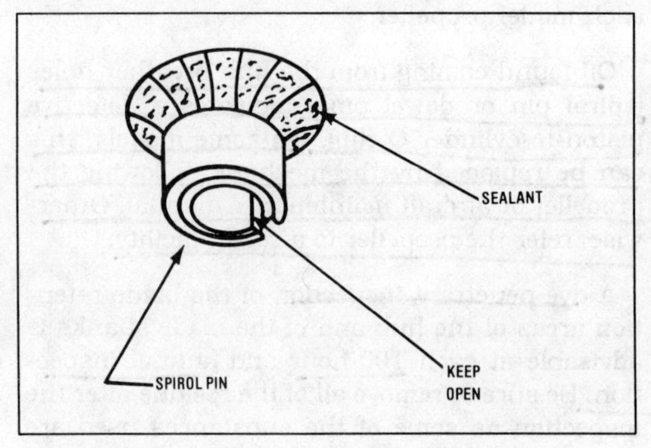

Figure 10-31. Seal around the hub spirol pin to prevent water from entering the propeller hub.

Remember two basic differences — a loss of oil pressure will allow the propeller to go to the low blade angle stop and the governor is spring loaded to the high RPM setting in case the cockpit control should break.

D. Hartzell Propeller System

The Hartzell constant-speed propeller systems are used in modern general aviation aircraft and share the market with McCauley. Hartzell systems are used extensively on Piper aircraft and on many other designs.

1. Propellers

Hartzell produces two styles of constant-speed propellers — a *Steel Hub* propeller and a *Compact* model. Steel Hub propellers are identified by the exposed operating mechanism, while Compact models have the pitch-changing mechanism encased in the hub assembly.

Some models of the Steel Hub propellers use oil pressure to decrease blade angle and the centrifugal force on the counterweights to increase blade angle. Other models of the Steel Hub propellers use centrifugal twisting moment to decrease blade angle and oil pressure to increase blade angle.

Hartzell Steel Hub propellers use a steel spider as the central component. Bearing assemblies and aluminum blades are placed on the spider arms and are held in place by two-piece steel clamps. A steel cylinder is screwed onto the front of the spider and an aluminum piston is placed over the

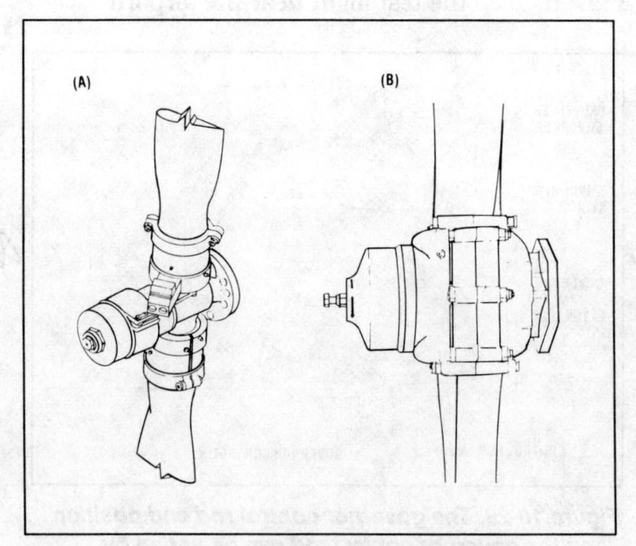

Figure 10-32A. Hartzell Steel Hub propeller.
Figure 10-32B. Hartzell Compact propeller.

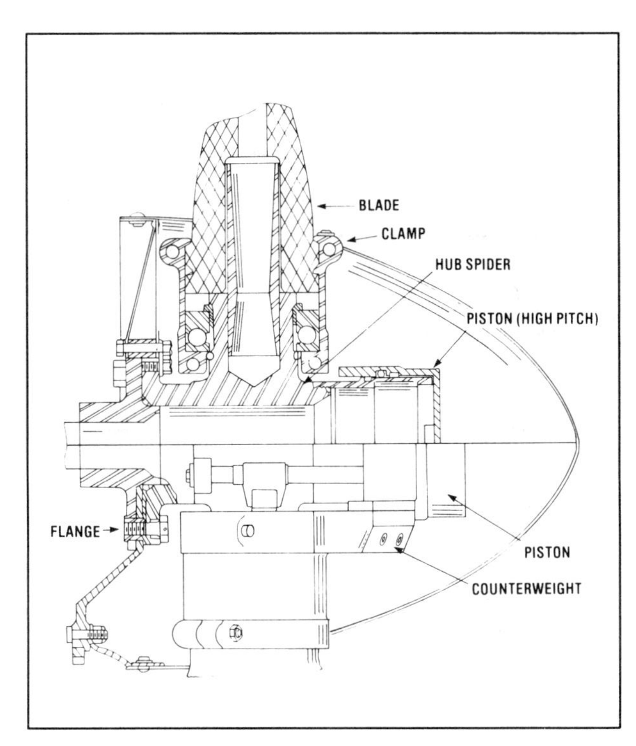

Figure 10-33. Cutaway view of a Hartzell Steel Hub propeller.

cylinder. The piston is connected to the blade clamps by steel link rods.

During operation oil pressure is directed to the propeller piston through the engine crankshaft and causes a change in blade angle. Counterweighted models use oil pressure to decrease the blade angle and centrifugal force on the counter weights to increase the angle. Non-counterweighted models use oil pressure to increase blade angle and centrifugal twisting moment to decrease the angle.

The Steel Hub propellers may be designed to mount on a flanged or a splined crankshaft.

Hartzell Compact propellers use aluminum blades mounted in an aluminum hub. The hub is held together with bolts and contains the pitch-changing mechanism of the propeller consisting of a piston, piston rod, and actuating links.

The Compact propellers use governor oil pressure to increase blade angle and the centrifugal twisting moment acting on the propeller blades to decrease blade angle. Some models also use counterweights to aid in increasing the blade angle.

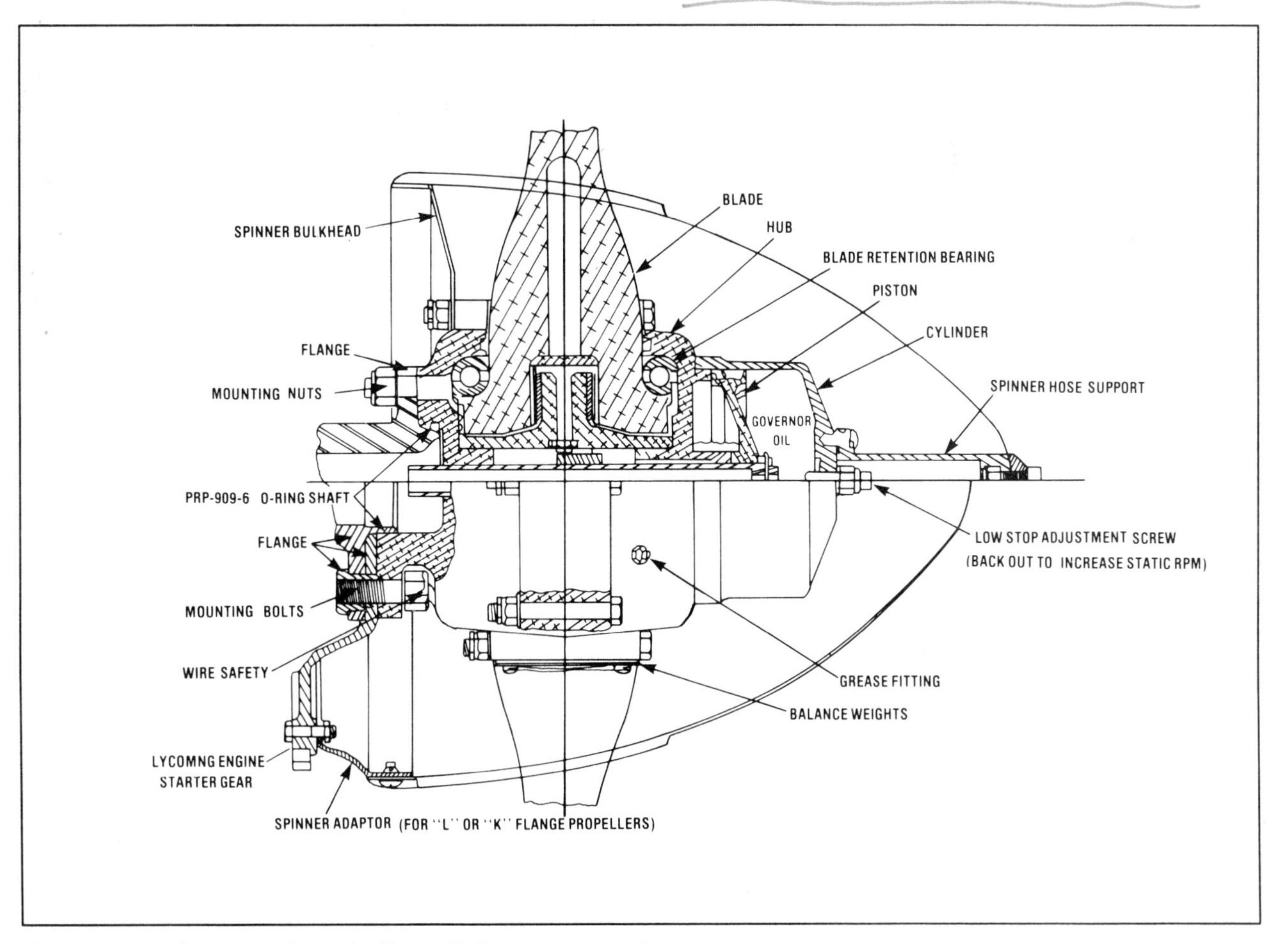

Figure 10-34. Cutaway view of a Hartzell Compact propeller.

The Hartzell propeller designation system is the same for Steel Hub and Compact propeller models as shown in Figure 10-35. It should be noted that a greater variety of designators for each component of the designation system exists than are illustrated here. A full listing of all of the variations in listings would not be practical for this text.

2. Governors

Hartzell governors are reworked Hamilton-Standard governors and use the designation system shown in Figure 10-36. Many Hartzell installations use a Woodward brand governor, which is similar to the McCauley governor in appearance and operation.

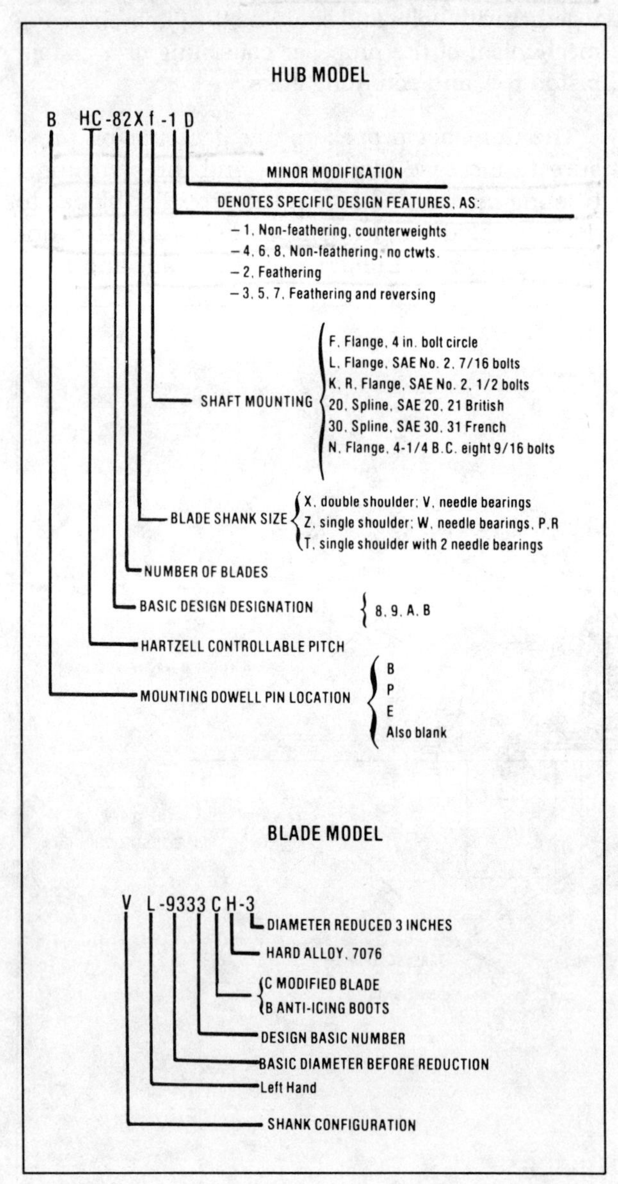

Figure 10-35. Hartzell propeller and blade designation system.

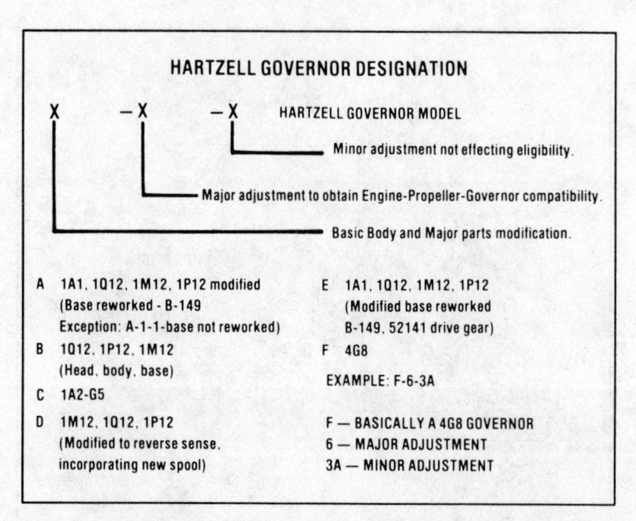

Figure 10-36. Hartzell governor designation system.

3. Installation And Adjustment

Hartzell propellers are installed following the practices discussed in Chapter V and covered in the section on McCauley constant-speed propellers.

The Steel Hub propellers can be adjusted for the desired low blade angle by loosening the hub clamps and rotating the blades in the clamps until the desired blade angle is obtained. The clamps are then retorqued and safetied. This also changes the high blade angle since the range between high and low blade angle is fixed by the range of the piston movement.

The low pitch setting of the Compact propellers can be adjusted by loosening the lock nut on the adjusting screw on the hub cylinder and rotating the screw in to increase the low blade angle or out to decrease the blade angle. When the desired angle is set, retighten the lock nut.

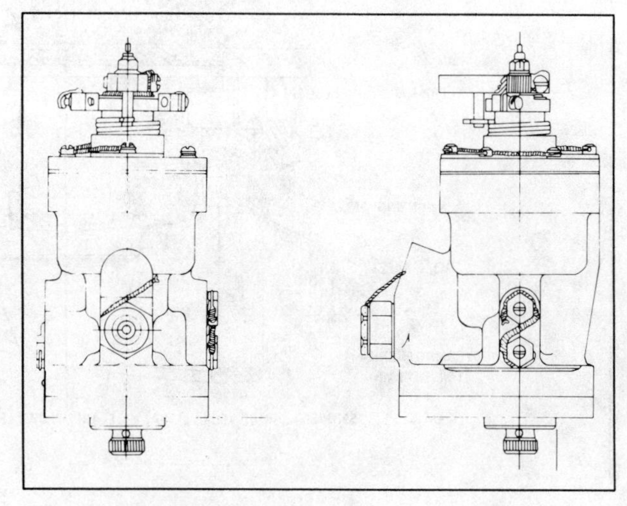

Figure 10-37. Woodward governors are used with some Hartzell propeller installations.

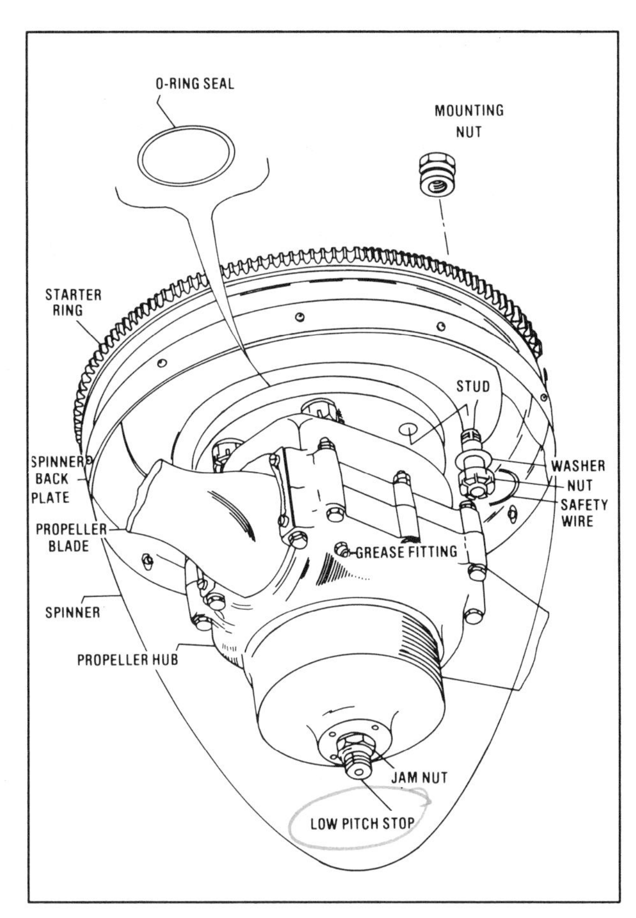

Figure 10-38. Hartzell Compact propeller installation on a Lycoming engine.

When changing the blade angles, always refer to the aircraft specifications and the propeller manufacturer's manual for instructions about specific propeller models.

The governors are installed and adjusted as has been discussed for the Hamilton-Standard and McCauley systems. Again, refer to the aircraft maintenance manual for information about specific installations.

4. Inspection, Maintenance And Repair

The inspection, maintenance, and repair of the Hartzell constant-speed propeller systems is basically the same as for systems previously discussed. However, special care should be taken when lubricating the propeller blades to prevent damage to the blade seals.

Before lubricating a blade, remove one of the two zerk fittings for the blade and grease the blade through the remaining zerk fitting. This will prevent any pressure from building up in the blade grease chamber and prevent damage to the blade seals. When grease comes out of the vacant zerk fitting hole, sufficient grease has been applied to the blade. Reinstall the zerk fitting, replace the protective cap and safety it.

Overhaul intervals are generally specified by the manufacturer, and the most recent service information should be consulted.

5. Troubleshooting

The troubleshooting procedures and corrections previously discussed are applicable to the Hartzell system. The principle addition is the determination of the source of a grease leak.

A grease leak is readily noticeable and the cause should be determined and corrected as soon as possible. The most common causes of grease leakage are loose or missing zerk fittings, defective zerk fittings, loose blade clamps, defective blade clamp seals, and over-lubrication of the blades.

If the zerk fitting is loose, missing, or defective, it should be tightened or replaced as appropriate. Loose blade clamps should be torqued to the specified value for the particular propeller model in question and resafetied. Check to be sure that the blade angle has not changed. Defective blade clamp seals and damaged grease seals should be replaced by an overhaul facility.

QUESTIONS:

1. *Which component of a constant-speed system is considered an RPM sensing device?*
2. *What is the purpose of the governor pilot valve?*
3. *What components of the governor opposes the force of the speeder spring?*
4. *During constant-speed operation, if the cockpit propeller control is moved forward, will blade angle increase or decrease?*
5. *If an aircraft with a constant-speed is in cruising flight and the aircraft enters a climb, will propeller blade angle increase or decrease?*
6. *In cruising flight, if the propeller control is pulled aft, what change in manifold pressure will occur?*
7. *When rotating a governor driveshaft by hand during the pre-installation inspection, a smooth resistance to rotation is felt. What does this indicate and what maintenance action must be taken?*
8. *What is the position of the screen on the governor mounting gasket during installation?*

9. What is an acceptable amount of cockpit control cushion when rigging the governor?

10. What is the purpose of a governor balance spring?

11. What would be the result of sludge building up in the governor?

12. Presuming that all components are adjusted correctly, what will cause momentary overspeeding of the propeller during takeoff?

13. What forces are used to change the blade angle of a McCauley constant-speed propeller?

14. What is the purpose of the oil-filled hubs used in some McCauley propeller designs?

15. How many times may the fiber lock nuts be used to attach the propeller to the crankshaft?

16. Where is an oil leak most likely to occur on a newly installed McCauley propeller?

17. If oil is found coming from the dowel pin breather hole in a McCauley propeller, what is a likely cause?

18. What is the most apparent difference between a Hartzell Compact propeller and a Steel Hub propeller?

19. What forces are used to operate a Hartzell Compact propeller?

20. How is the low blade angle of a Hartzell Steel Hub propeller adjusted?

21. What precaution should be taken before greasing a Hartzell propeller blade?

22. What will cause grease to leak from the Hartzell propeller blades?

Chapter XI
Feathering Propeller Systems

Feathering propellers are used on all modern multi-engine aircraft and on all but a few vintage multi-engine airplanes. The primary purpose of a feathering propeller is to eliminate the drag created by a windmilling propeller when an engine fails. There are no current production single-engine aircraft equipped with feathering propellers other than a few special purpose airplanes using turboprop installations (such as the PT6-powered Thrush Commander agricultural aircraft).

Feathering propeller systems are constant-speed systems with the additional capability of feathering the propeller. This means they have the ability to rotate the propeller blades to an approximate 90-degree blade angle. The constant-speed controls and operational events covered in Chapter X apply to the feathering propeller system. The cockpit propeller control lever incorporates an additional range of movement to allow feathering the propeller or a separate push-button control may be used to operate the feathering mechanism.

Feathering operations are independent of constant-speed operations and can override the constant-speed operation to feather the propeller at any time. The engine does not have to be developing power and in some systems the engine does not

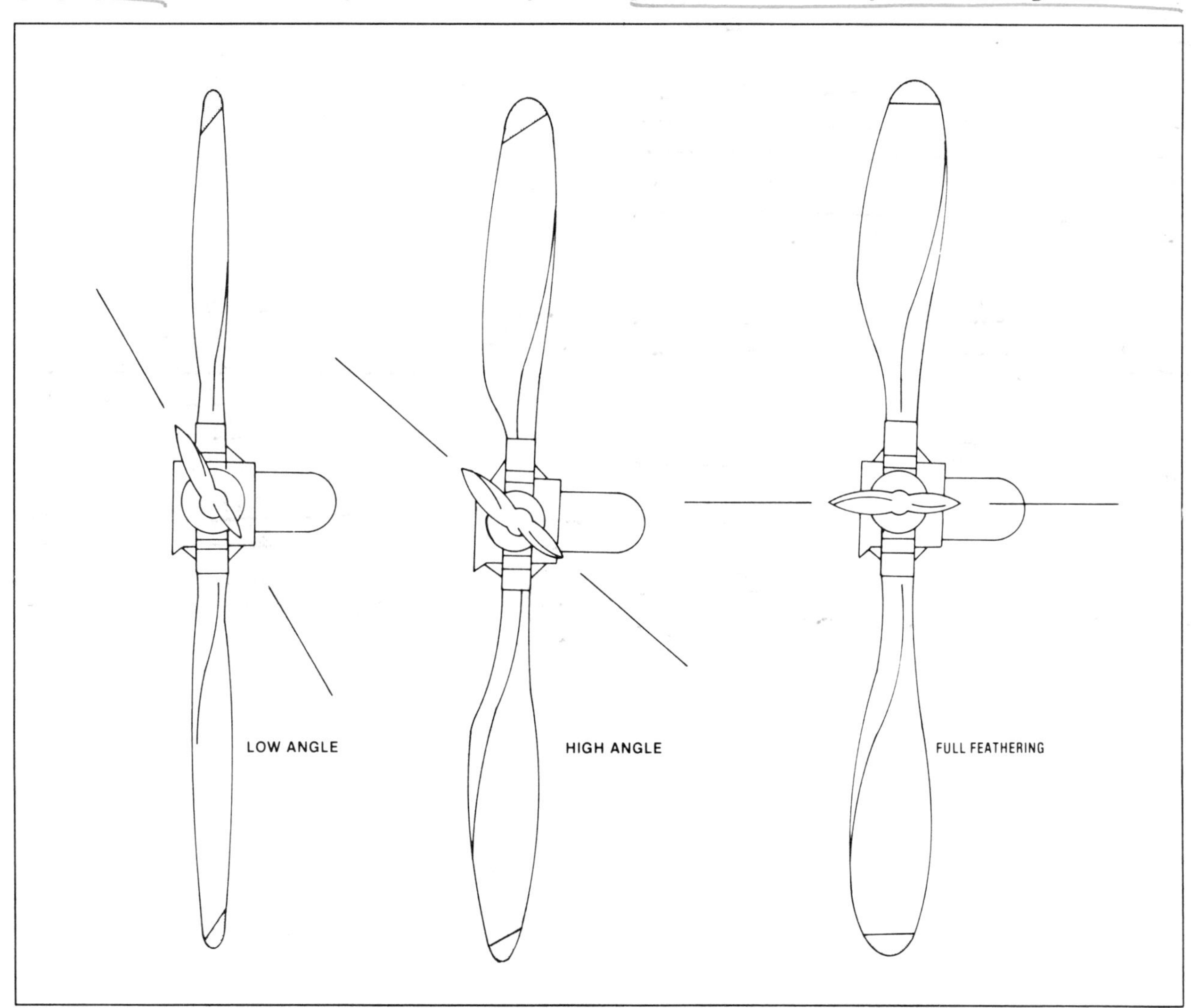

Figure 11-1. When the propeller feathers, the blades are about 90 degrees to the plane of rotation.

have to be rotating to feather the propeller. In short, propellers are feathered by forces which are totally independent of engine operation.

It should be noted that when the propeller is feathered, the engine stops rotating.

A. McCauley Feathering System

The McCauley feathering system is used on all current production piston-engined Cessna twins and on many twin-engined Beechcraft. The system incorporates a feathering propeller and governor, cockpit control levers which control both the constant-speed and feathering operations of the system, and unfeathering ***accumulators*** as optional equipment.

1. Propellers

The outward appearance of McCauley feathering propellers is similar to that of the constant-speed propellers except for the longer cylinder, which allows the greater blade angle range, and the addition of counterweights. The propeller is designed to use oil pressure from the governor to decrease blade angle while the force of springs and counterweights is used to increase blade angle and to feather the propeller. NOTE: Both threaded and threadless blade designs are used.

The propeller is spring-loaded and counterweighted to the feather position at all times so that if oil pressure is lost, the propeller will automatically feather. To prevent the propeller from feathering when the engine is stopped on the ground, a spring-loaded latch mechanism engages at some low RPM (for example 900 RPM). This prevents excessive load on the starter and engine system when starting the engine. Three different latch mechanisms have been used with the McCauley feathering propellers — the inertial latch, the pressure latch, and the centrifugal latch. Since the centrifugal latch has proved to be the best system, most propellers either have been converted to this style or will be converted during the next overhaul.

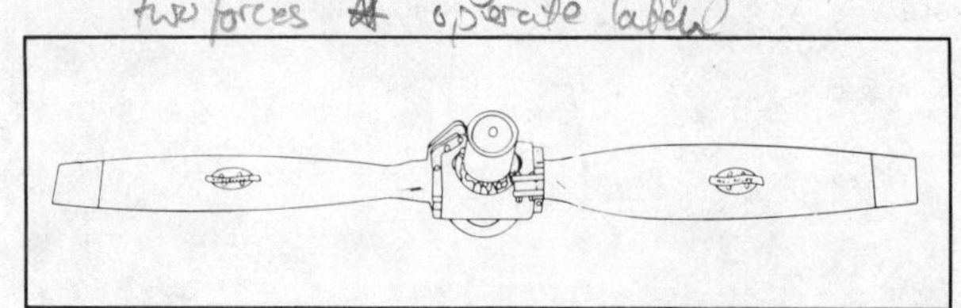

Figure 11-2. A McCauley feathering propeller. Note the cylinder size and blade counterweights and compare with the constant-speed model of the McCauley propeller.

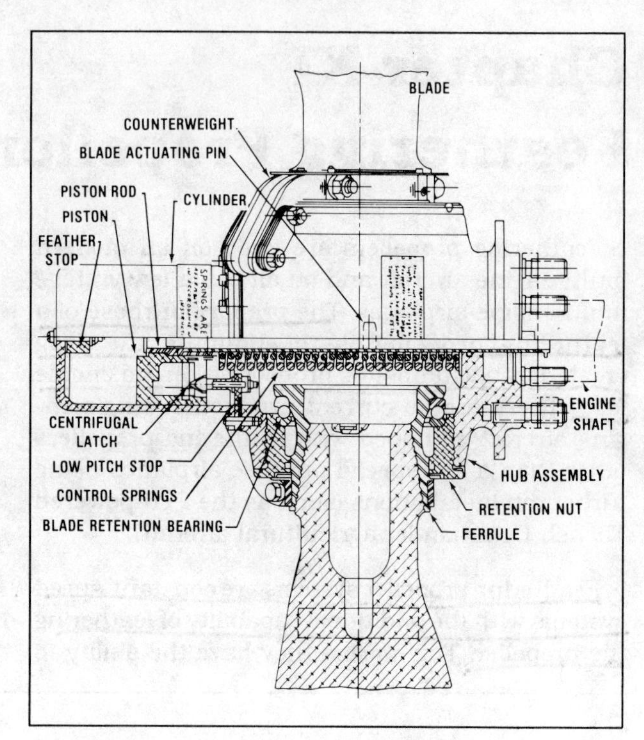

Figure 11-3. Cutaway view of a McCauley feathering propeller.

The propeller designation system is the same as for the McCauley constant-speed propellers.

2. Governors

McCauley feathering governors are basically the same as constant-speed governors except that the governor directs oil pressure to the propeller to decrease blade angle and releases oil from the propeller to increase blade angle.

Feathering governors incorporate a lift rod connected to the speeder rack which will mechanically lift the pilot valve, releasing the oil from the propeller when the cockpit control lever is moved to the feather position. This occurs when the cockpit control is pulled full aft. This action may take place

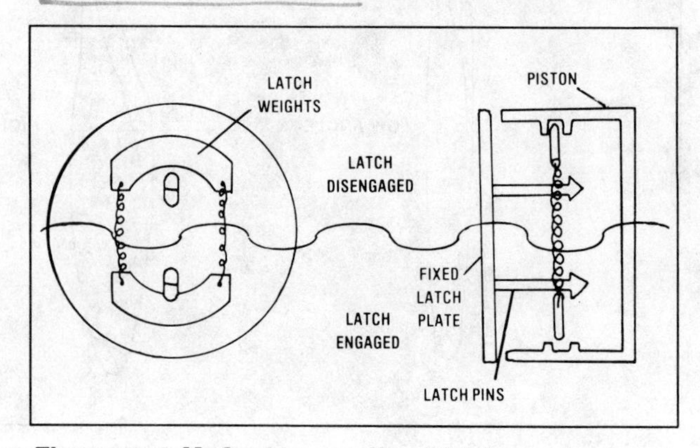

Figure 11-4. McCauley centrifugal latch mechanism.

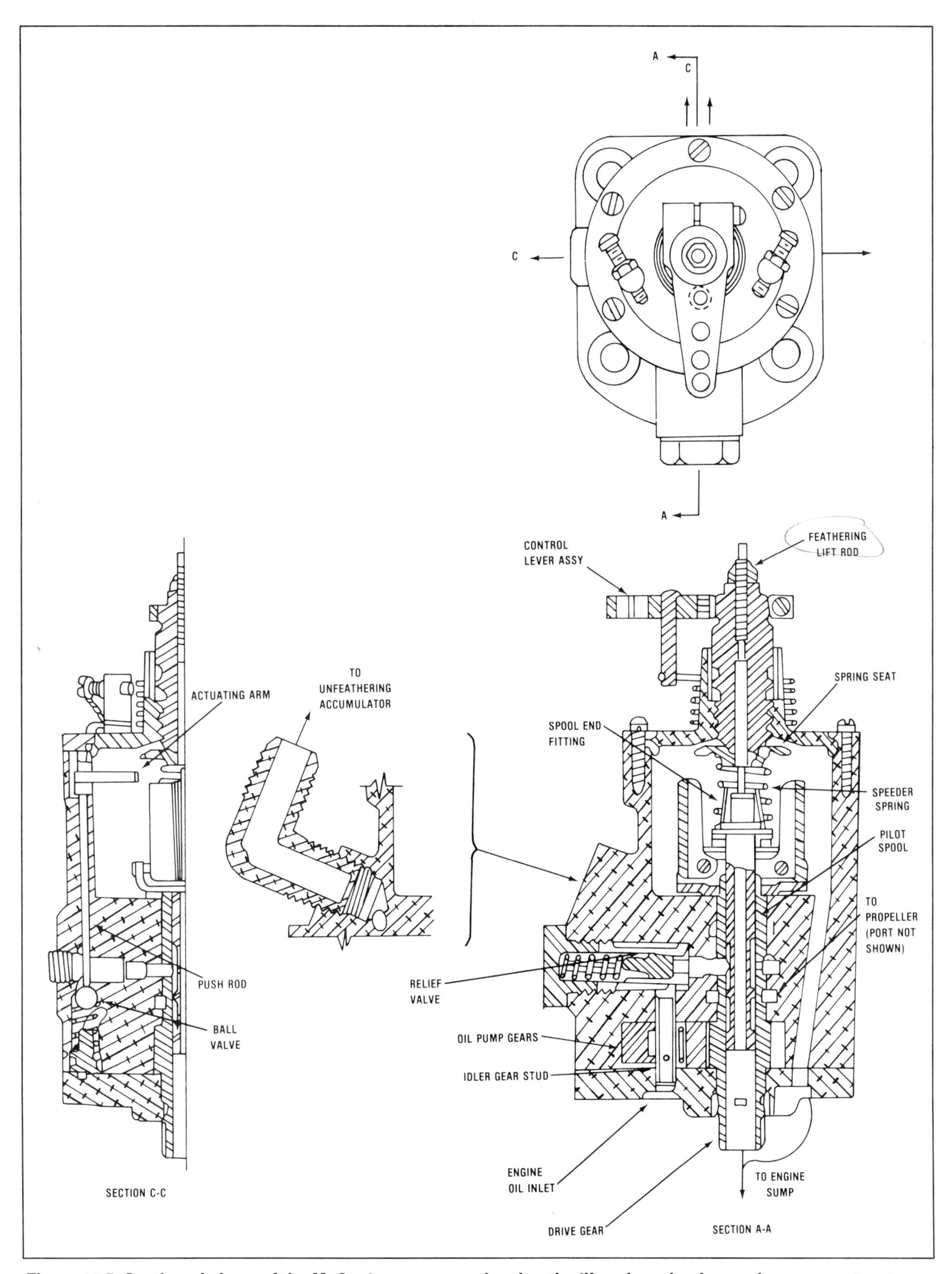

Figure 11-5. Sectioned views of the McCauley governor showing the lift rod mechanism and governor structure with and without accumulator passages.

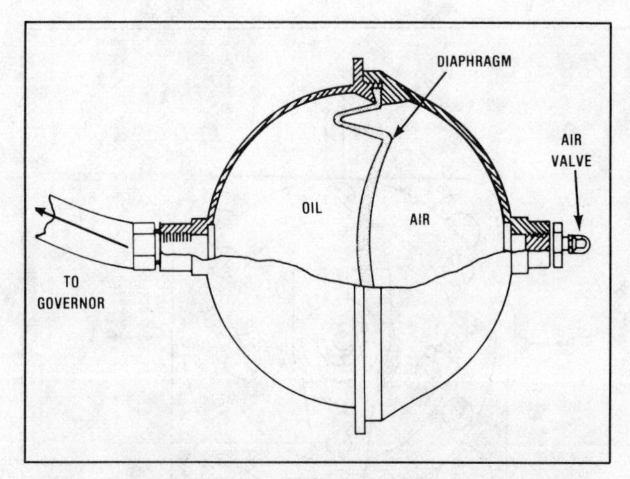

Figure 11-6. Cross section of a ball-type accumulator.

at any time and overrides the position of the flyweights or the speeder spring.

If the feathering system includes an accumulator, a ball-check valve will be incorporated in the governor design. This allows the accumulator to be charged during normal operation and prevents it from discharging during feathering. The ball check valve is released by a push rod when the propeller control is moved forward to the high RPM position. At this time, the stored oil pressure in the accumulator will be released to unfeather the propeller.

The governor designation system is the same as for the McCauley constant-speed governors.

3. Accumulator

The unfeathering accumulator is in the shape of a ball or a cylinder. It contains a diaphragm or piston which is used to separate an air charge from an oil charge. The air charge is approximately 100 psi and uses nitrogen or dry air. The oil side of the accumulator is charged by the governor through a flexible line and stores a charge of oil at governor pressure (290 psi). The accumulator is usually installed in the engine compartment.

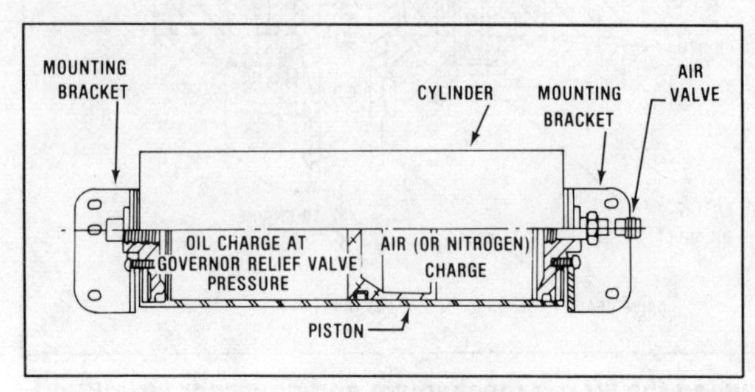

Figure 11-7. Cutaway view of a McCauley accumulator.

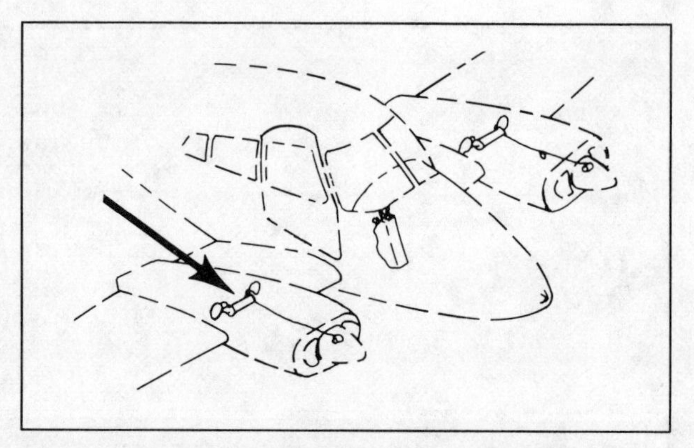

Figure 11-8. Typical location of an accumulator.

4. System Operation

The constant-speed operation of the McCauley feathering type propeller system is exactly the same as the McCauley constant-speed system except for the change in oil flow direction.

Feathering of the McCauley system is achieved by moving the cockpit propeller control lever full aft. When this occurs, the governor lever arm is pulled to the low RPM stop. This causes the RPM lift rod in the governor to raise the pilot valve and release oil pressure from the propeller. Without oil pressure in the propeller, it goes to the feather angle by a combination of spring force and centrifugal force on the blade counterweights. When the propeller blades have fully feathered, engine rotation stops. If an accumulator is used, the ball check valve in the governor holds a charge of oil pressure in the accumulator.

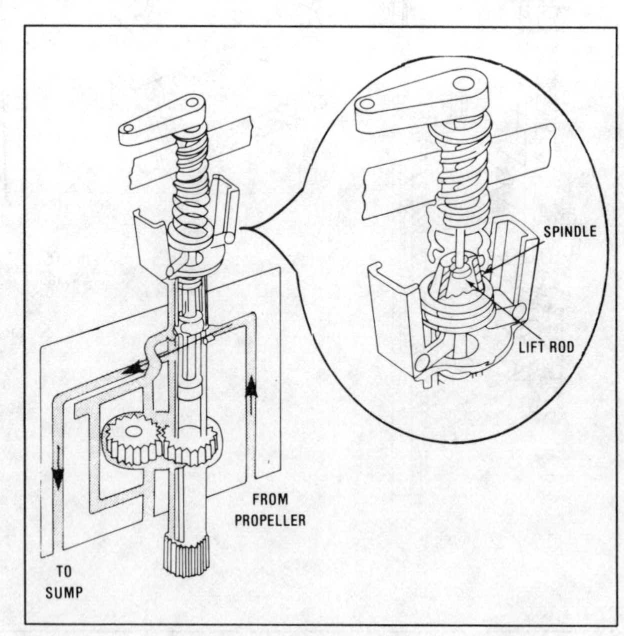

Figure 11-9. Action of the governor lift rod during feathering.

To unfeather the propeller without an accumulator, the cockpit control is moved forward and the engine must be rotated by the starter. This allows governor oil pressure to be developed to overcome the force of the springs and move the blades to a lower angle. As this blade angle decreases, the propeller will start to windmill and help complete the unfeathering operation.

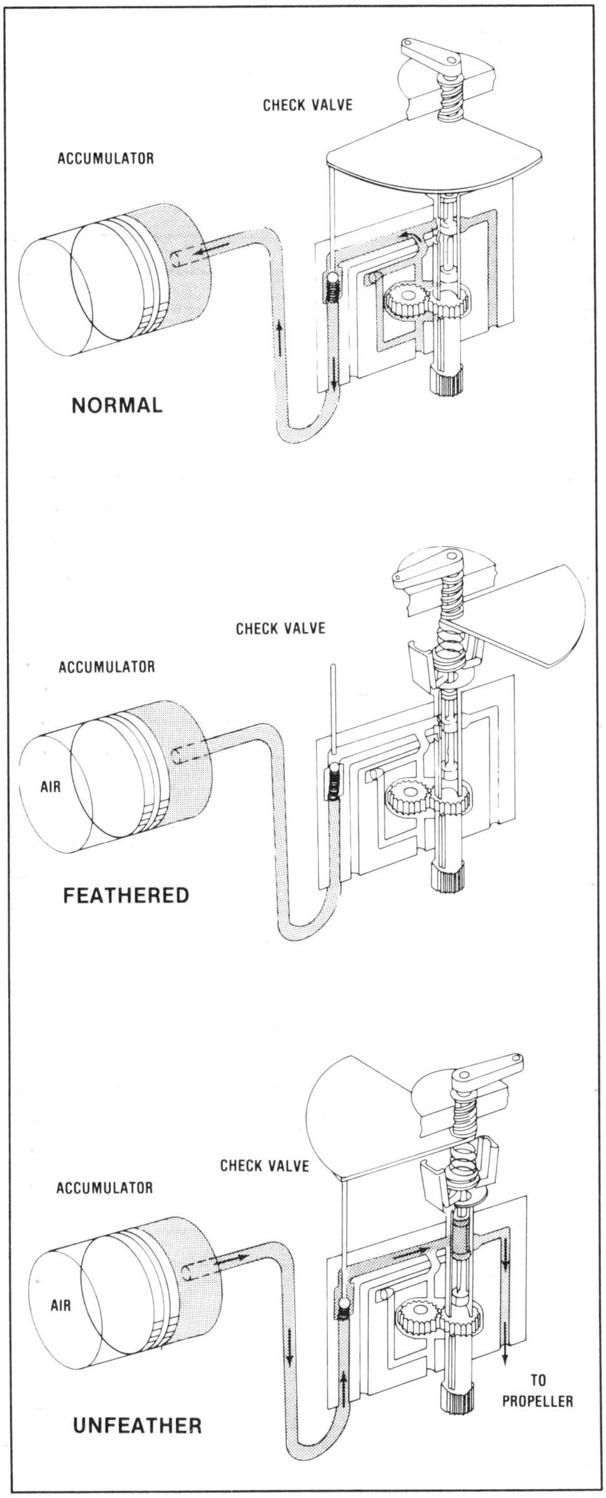

Figure 11-10. Action of the governor push rod, ball check valve and accumulator during system operation.

If an accumulator is used to unfeather the propeller, the governor ball check valve is merely opened by the push rod in the governor when the cockpit control is moved forward. This allows the oil pressure in the accumulator to flow through the governor to the propeller cylinder and force the blades to a lower angle. NOTE: The engine does not have to be attempting to restart for the accumulator to unfeather the propeller.

When stopping the engine on the ground, the engine should be idled to about 800 RPM with the cockpit propeller control in the full forward position. When the blades are at this low blade angle and the engine is shut down, they will tend to feather due to the loss of oil pressure. This action is prevented by a spring force which engages the latch mechanism. This force is greater than the centrifugal force on the latch plates at idle; consequently, the blades are held by latch plates at an angle a few degrees above the low blade angle as they move toward feather (see Figure 11-4).

When the engine is started on the ground, the cockpit control should always be full forward. This causes the blade angle to decrease as governor oil pressure is generated, and the blades will move to the low blade stop from the latch angle. As RPM increases, these latch plates move outward by centrifugal force allowing propeller blades to be free to move through their full range of travel.

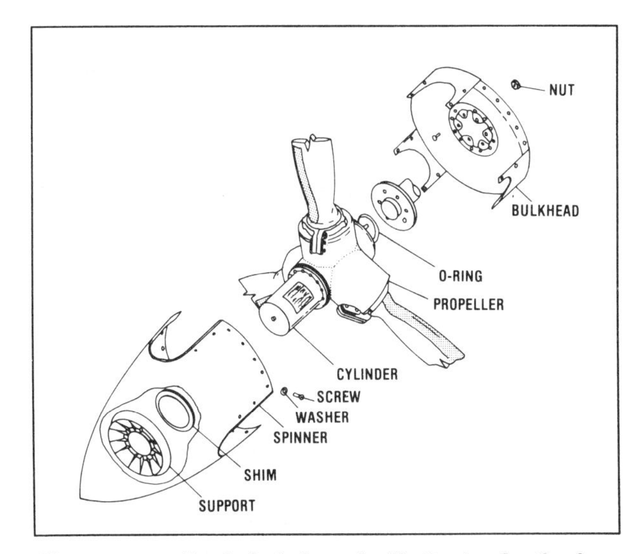

Figure 11-11. Exploded view of a McCauley feathering propeller installation.

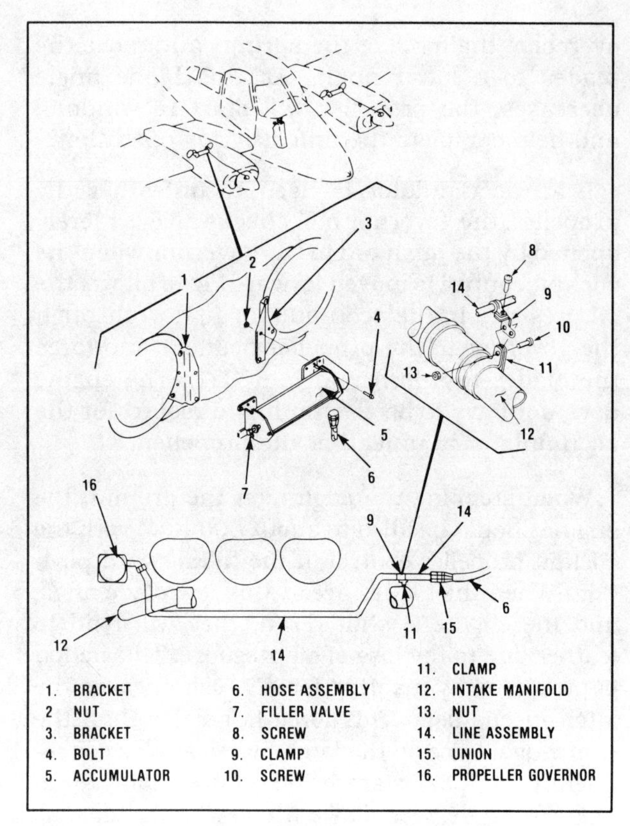

Figure 11-12. Typical accumulator installation.

5. Installation And Adjustments

The installation and adjustment of the feathering propeller and governor are basically the same as in the constant-speed models. If an accumulator is installed, it is usually installed in the engine compartment in accordance with standard airframe practices. It is connected to the governor with flexible hose similar to those used in hydraulic systems (Figure 11-12).

6. Inspection, Maintenance And Repair

Inspection of the McCauley feathering propeller is the same as that of the constant-speed propellers. However, special attention should be given to checking the counterweights for slippage and security.

The governor should be inspected in the same manner as the constant-speed governors. If an accumulator is used, be sure to check the fittings for security and leaks.

The accumulator should be inspected for security of mounting, proper air charge, and leaks. Be sure to check the lines from the accumulator to the governor for condition, wear, security and leaks.

The propeller should be overhauled at 1,200 hours or at engine overhaul. The governor should be overhauled at 800 hours or at engine overhaul. The accumulator should be overhauled at engine overhaul or as specified by the manufacturer.

7. Troubleshooting

In addition to the troubleshooting procedures discussed for constant-speed systems, the feathering system may have some additional operating difficulties.

If the propeller will not feather in flight, the problem may be that the governor low RPM stop is set too high preventing the lift rod from raising the pilot valve. It may also be caused by the cockpit controls being rigged improperly or restricted, preventing full movement of the governor lever arm.

If the propeller high blade angle is not correctly set, the propeller may continue to windmill after the propeller feathers. Refer the propeller to an overhaul facility to adjust the high pitch stop.

If the propeller fails to unfeather in flight and an accumulator system is not used, the problem may be that the starter cannot generate sufficient RPM to restart the engine or develop sufficient oil pressure in the governor to unfeather the propeller. This is a common problem is some older aircraft and involves a change in pilot technique and/or the addition of accumulators. Some aircraft can be unfeathered more easily if the aircraft is placed in a shallow dive to increase airspeed, then rotating the engine with the starter.

If the propeller fails to unfeather with an accumulator installed, one of the following problems may be the cause — a loss of air pressure or insufficient air pressure in the accumulator, oil leaking from the accumulator or flex hose causing a reduction in the oil pressure, or a leak from the air side of the accumulator to the oil side resulting in a loss of air pressure. External air leaks can be located by pressurizing the system to normal operating pressure and checking for leaks with soapy water.

To locate the problem, simply follow these guidelines. Oil leaks are readily apparent. A loss of air charge to the oil side of the accumulator can be checked by draining the oil side of the accumulator and checking for air coming out of the oil line fitting. If air is leaking to the oil side of the accumulator, the accumulator should be overhauled.

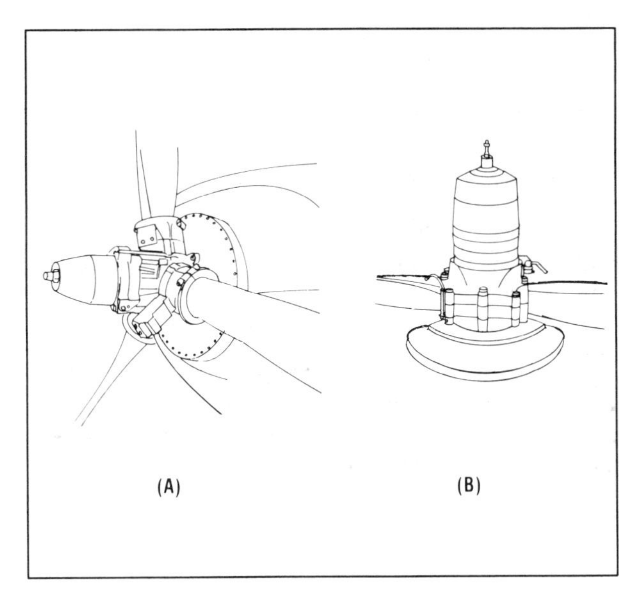

Figure 11-13A. Hartzell Steel Hub propeller.

Figure 11-13B. Hartzell Compact feathering propeller.

B. Hartzell Feathering System

Hartzell feathering systems are used on all current production Piper piston-driven twins, some Beechcraft and Aero Commander twins, and older Cessna twin-engine aircraft.

1. Propellers

Both Compact and Steel Hub propeller designs are used for Hartzell feathering propellers.

Compact designs use governor oil pressure to decrease blade angle and air pressure in the propeller cylinder (and counterweights on some models) to feather the propeller. A latch stop (called the High Pitch Stop by Hartzell) is located inside the cylinder to hold the blades in a low blade angle when the engine is stopped on the ground. The latch mechanism is composed of springs and locking pins.

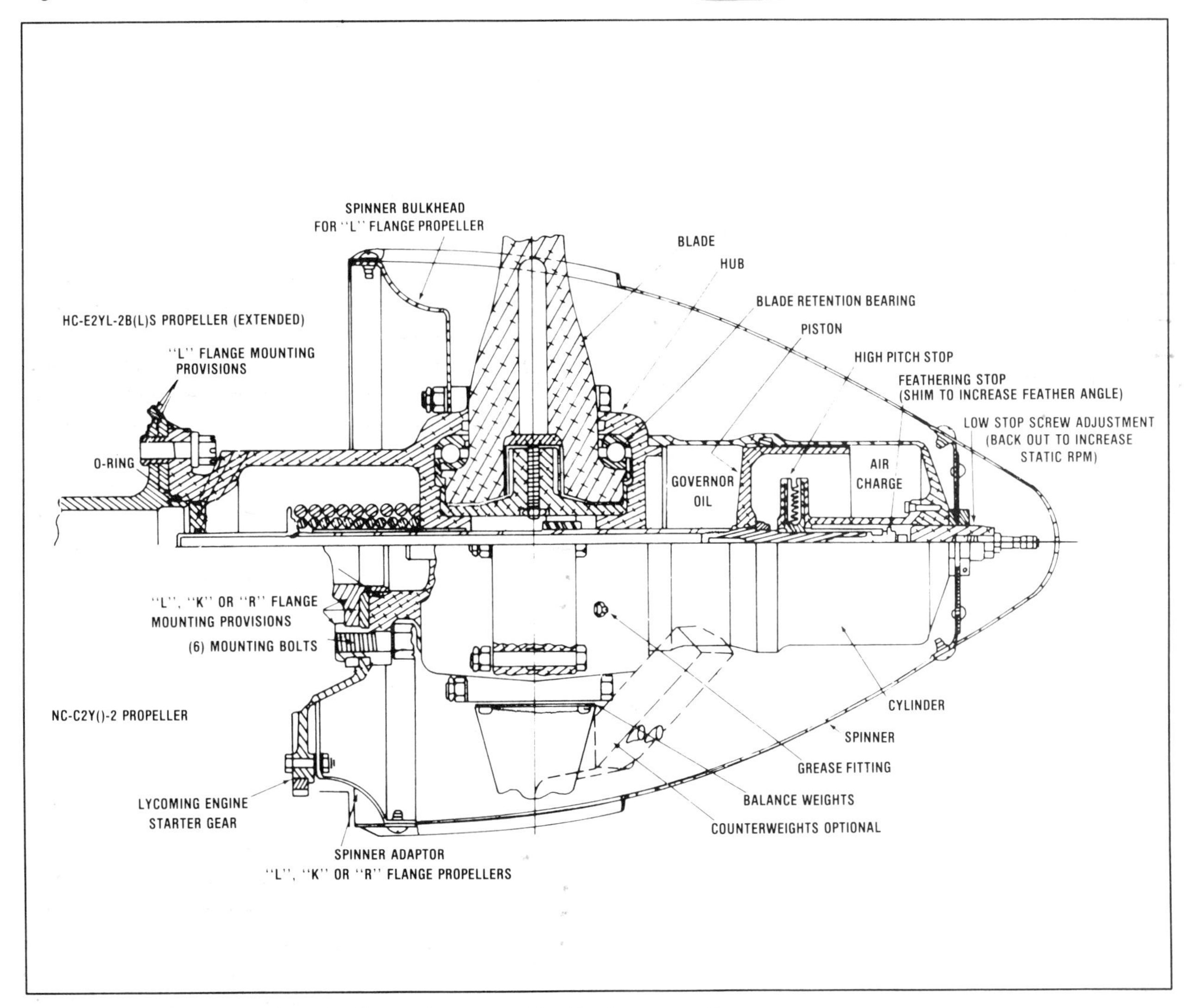

Figure 11-14. Cross section of a Hartzell Compact feathering propeller.

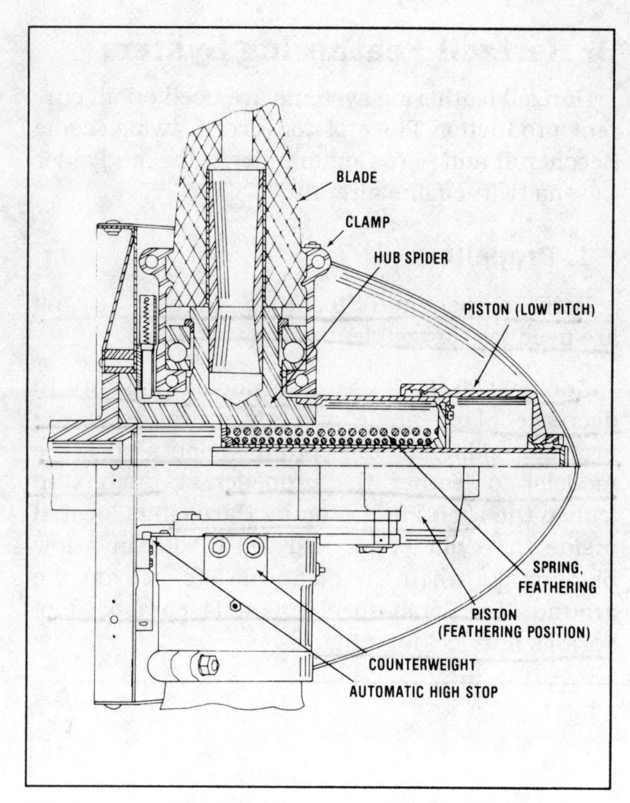

Figure 11-15. Cross section of a Hartzell Steel Hub feathering propeller.

The Hartzell Steel Hub feathering propeller uses oil pressure to decrease blade angle and a combination of springs and counterweights to increase the propeller blade angle. An external latch mechanism is used to prevent the propeller from feathering when the engine is stopped on the ground.

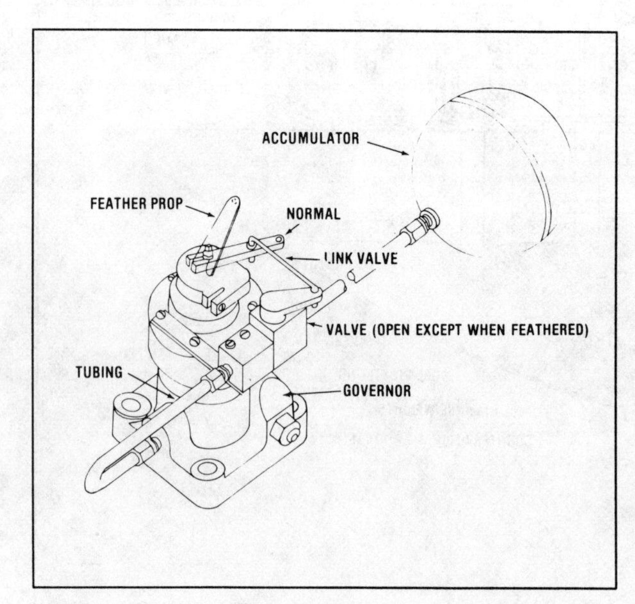

Figure 11-16. Governor with an external accumulator adapter.

The Hartzell feathering propeller designation system is the same as for the constant-speed models.

2. Governors

Hartzell feathering systems may use either Hartzell or Woodward governors for operation. The governors may incorporate an internal mechanism with a lift rod and accumulator oil passages and valves. Or, they may have an external adapter which contains a shutoff valve linked to the governor control arm to control the accumulator operation.

3. Accumulators

The information concerning accumulators that is covered in the section on McCauley feathering propellers is applicable to the Hartzell feathering system.

4. System Operation

The constant-speed operation of the Hartzell feathering propellers is the same as for the constant-speed models except for the change in direction of oil flow in some models.

When the Hartzell propellers are feathered, the cockpit control is moved full aft and the governor pilot valve is raised by the lift rod to release oil from the propeller. With the oil pressure released, a Steel Hub propeller will move to the feather position by the force of the counterweights and springs. The Compact models will go to feather by the force of the air pressure in the cylinder and the force of the backup spring, in some models, the force of the counterweights. The blades are held in feather by the spring force or air pressure.

When unfeathering the propeller in flight, the system relies on engine rotation by the starter to initiate the unfeathering operation unless an accumulator is used. The operation of the accumulator is the same as for the McCauley system.

When shutting the engine down after flight, the propeller cockpit control should be placed in full forward position while the engine is idling. This causes the spring in the latch mechanism to force the lock pin into a low ***pitch lock*** position and engage when the engine is shut down and the blades attempt to rotate toward feather.

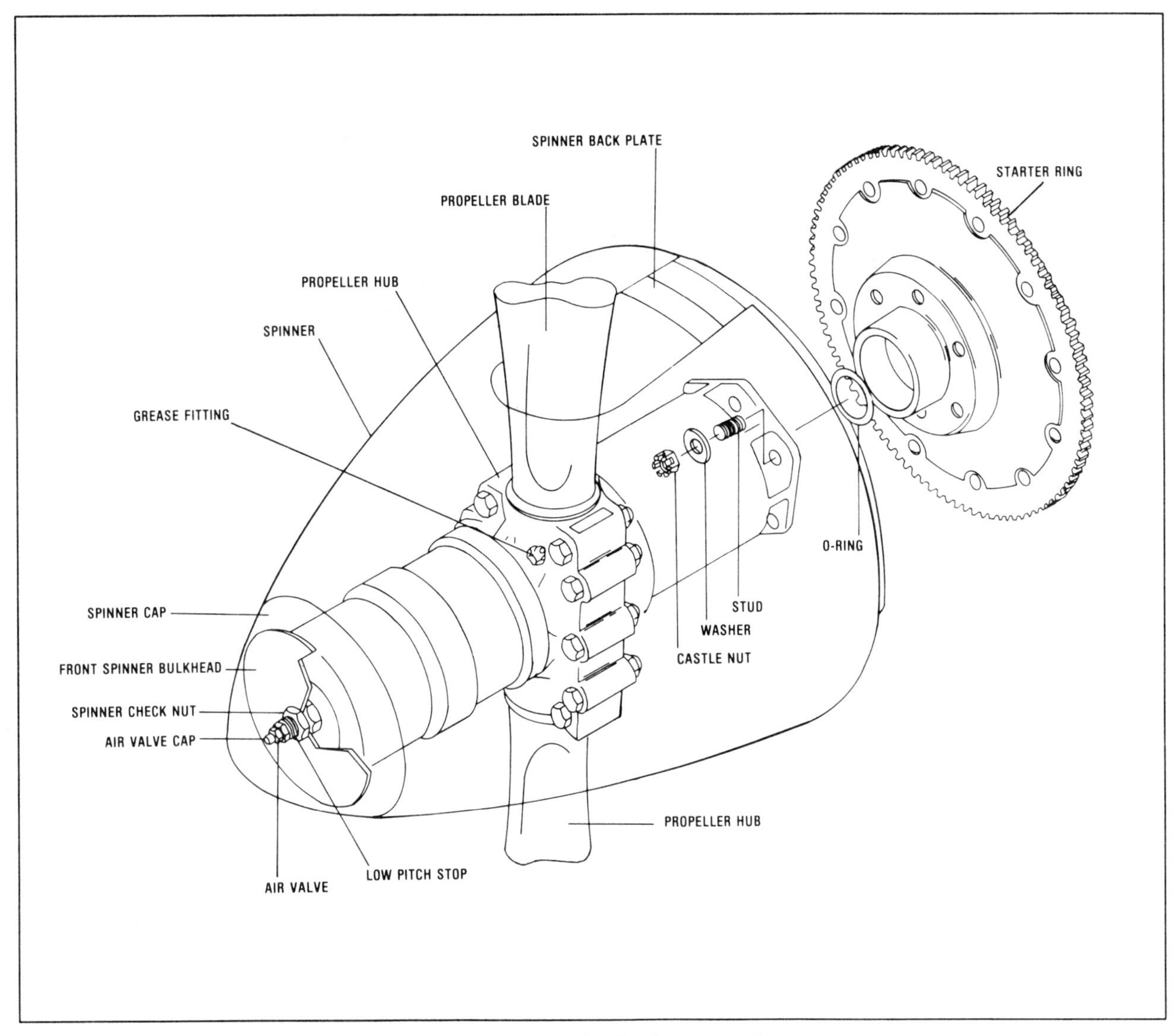

Figure 11-17. Typical installation of a Hartzell Compact feathering propeller.

5. Installation And Adjustments

Installation and adjustment of the propellers are the same as for the constant-speed models. If the blades are feathered, they can be rotated to the latch angle by placing a blade paddle on each blade and rotating the blades to the latch angle simultaneously.

The air pressure in the Compact model propellers should be checked and serviced as necessary after installation. The amount of air pressure is determined by using a chart similar to the one in Figure 11-18.

Blade latches may be supplied separate from the Steel Hub propeller and are simply bolted onto the outside of the propeller in accordance with the instructions relating to the particular model propeller.

Governor installation and adjustment is the same as for a constant-speed governor.

Accumulators are installed in accordance with the aircraft maintenance manual and are serviced with dry air or nitrogen to the value specified in the aircraft maintenance manual.

CHAMBER PRESSURE REQUIREMENTS WITH TEMPERATURE			
TEMP. °F	PRESS. (PSI)	TEMP. °F	PRESS. (PSI)
100	188	30	165
90	185	20	162
80	182	10	159
70	178	0	154
60	175	−10	152
50	172	−20	149
40	168	−30	146

Figure 11-18. Air pressure chart for a Hartzell Compact feathering propeller.

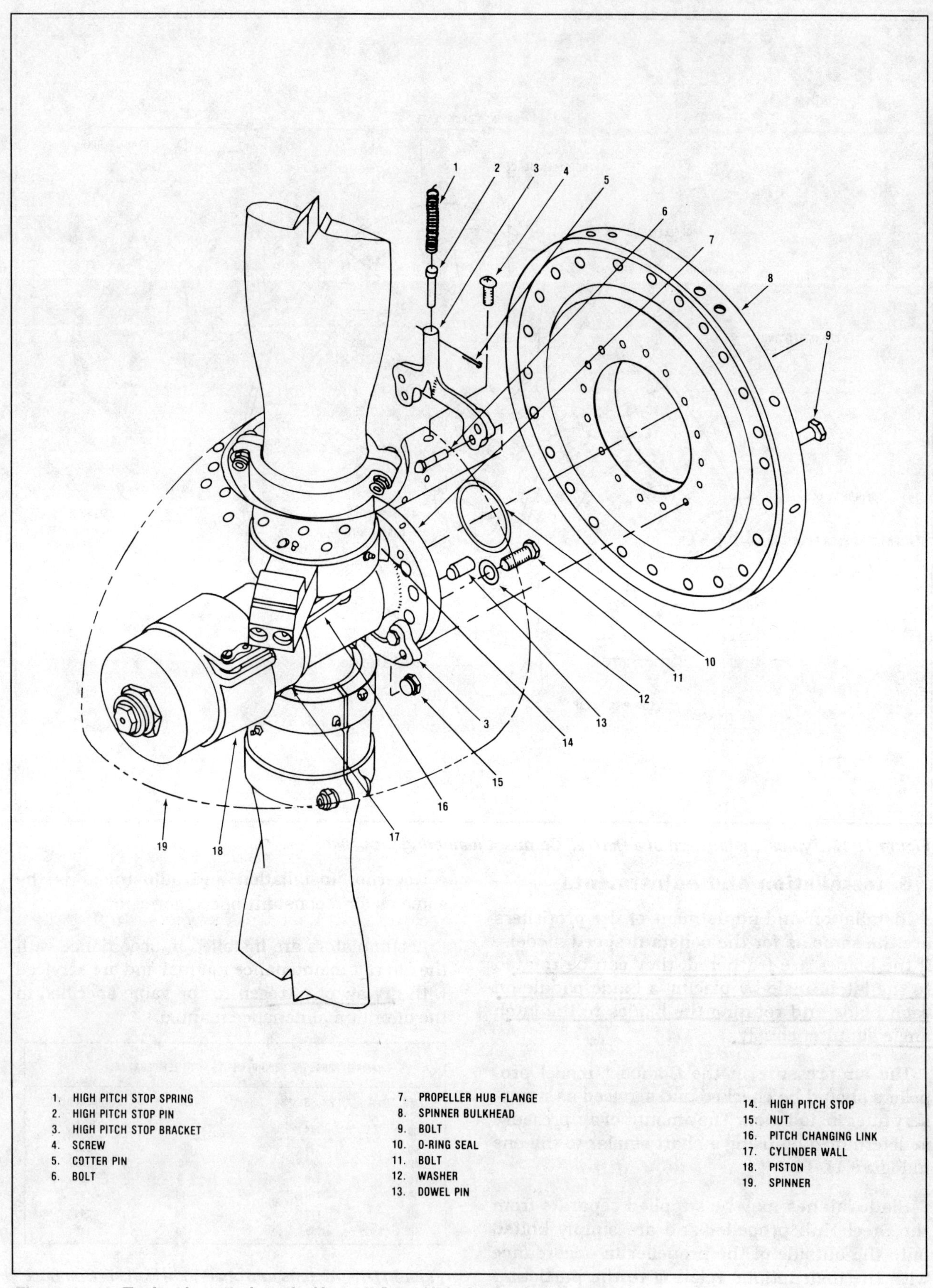

Figure 11-19. Typical installation of a Hartzell Steel Hub feathering propeller.

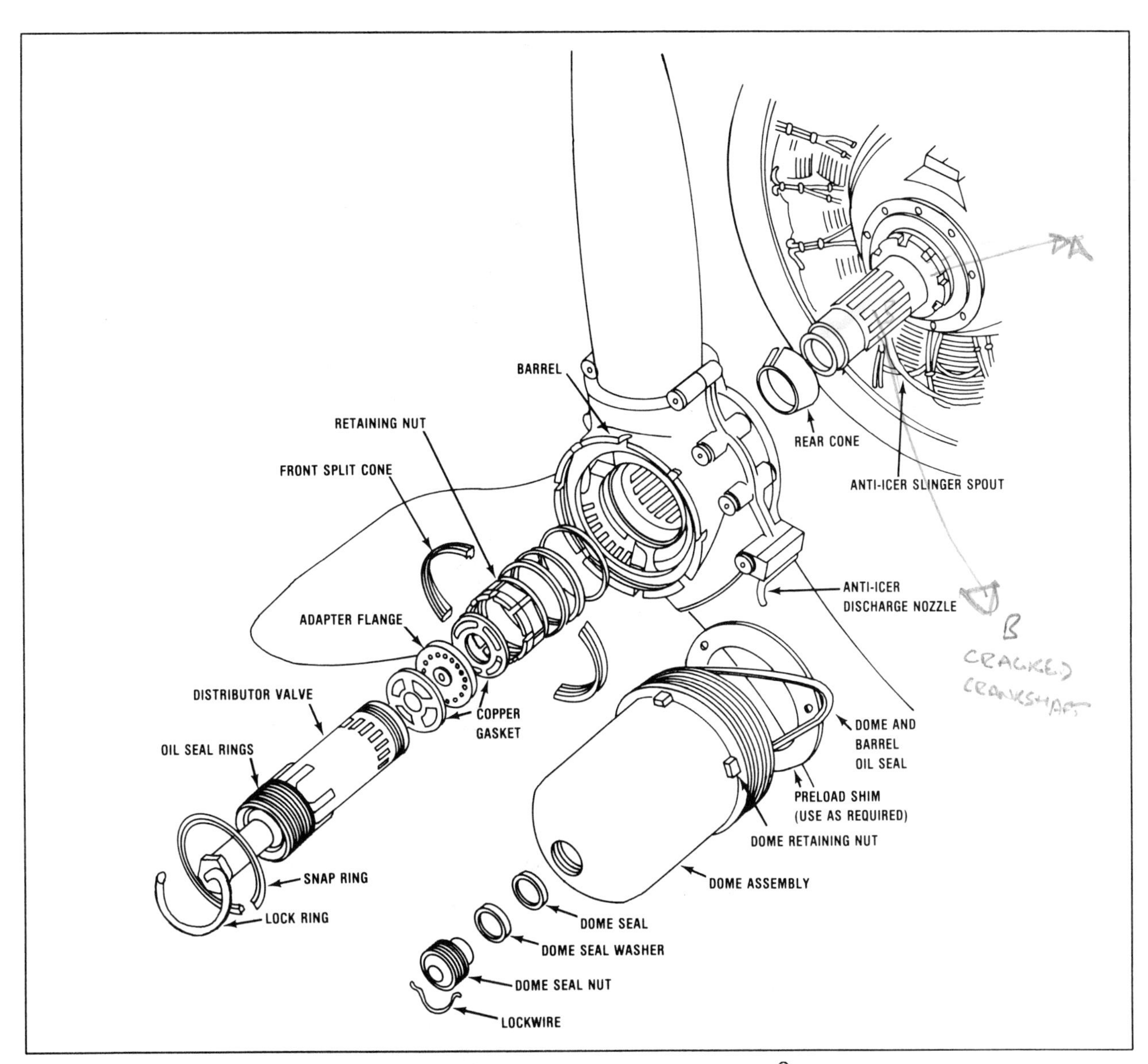

Figure 11-20. Exploded view of the Hamilton-Standard feathering Hydromatic® propeller.

6. Inspection, Maintenance And Repair

The inspection, maintenance, and repair of the Hartzell feathering propeller system is the same as described for other constant-speed and feathering systems. The Compact propeller air pressure should be checked at each 100-hour and annual inspection.

7. Troubleshooting

The system troubleshooting procedures are the same as described for other propeller systems.

If the air charge is too low in the Compact propeller, it may not feather or respond properly to constant-speed operation. The propeller may also have a tendency to overspeed or surge. Also, if the air charge is too great, the system may not reach full RPM and may feather when the engine is shutdown on the ground.

C. Hamilton-Standard Feathering System

The Hamilton-Standard feathering system is used on many medium and large transports such as the Beech 18, the Curtiss C-46, and the Douglas DC-4. The Hamilton-Standard design goes by the trade name of ***Hydromatic®*** indicating that the principal operating forces are liquid (oil pressure).

1. Propeller

The Hamilton-Standard Hydromatic® propeller is composed of three major assemblies — the hub

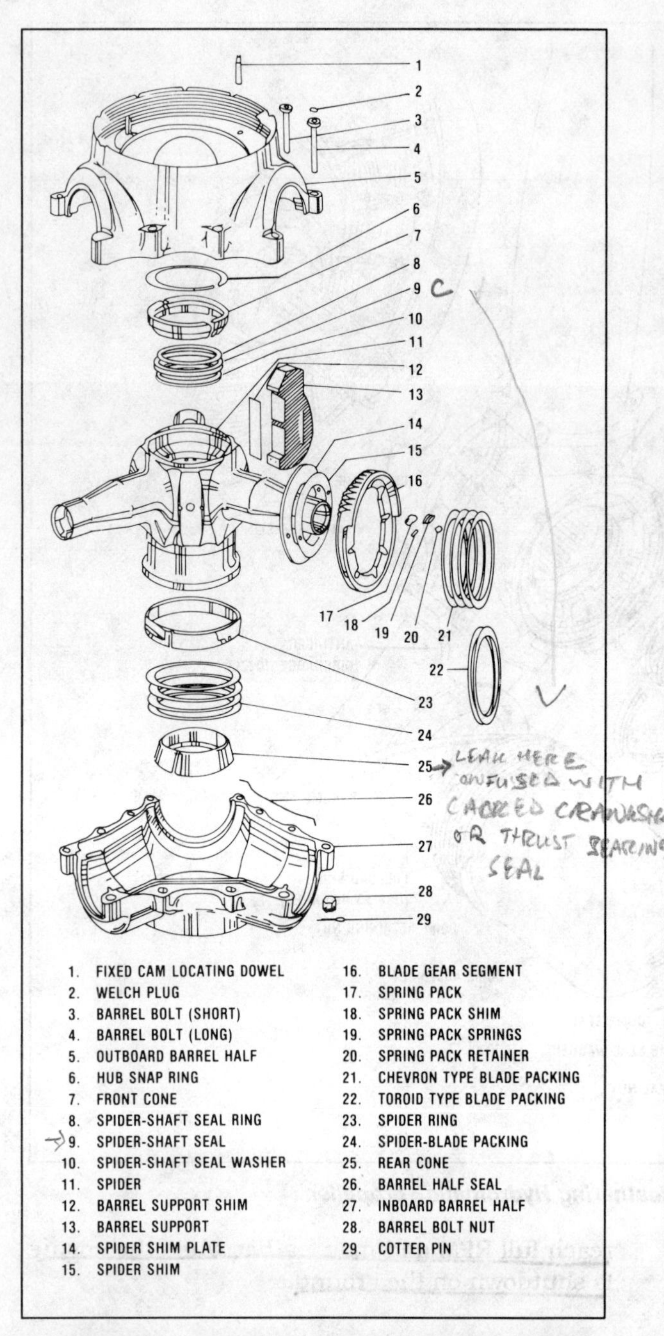

Figure 11-21. An exploded view of the barrel assembly without the blades. The blades fit on the arms of the spider.

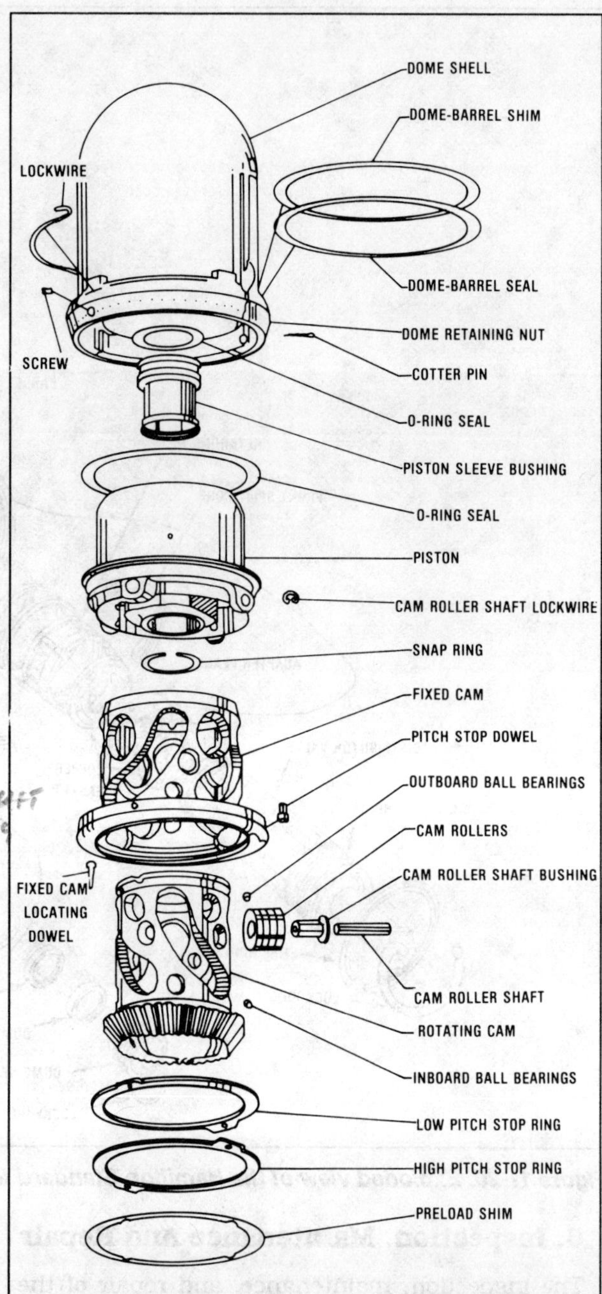

Figure 11-22. An exploded view of the dome assembly.

or barrel assembly, the ***dome assembly,*** and the distributor valve.

The barrel assembly contains the spider, blades, blade gear segments, barrel halves, and necessary support blocks, spacers, and bearings. Front and rear cones, a retaining nut, and a lock ring are used to install the barrel assembly on the crankshaft.

The dome assembly contains the pitch-changing mechanism of the propeller and includes the dome shell, a piston, a rotating cam cylinder, a stationary cam cylinder, and two pitch stop rings. The dome shell acts as the cylinder for the propeller piston. The piston is attached to the rotating and stationary cams by cam rollers which move in the slots in the cams. As the piston moves fore and aft in the dome shell, it rotates following the cam track in the stationary cam and causes the rotating cam to rotate. The gear on the end of the rotating cam meshes with the gear segment on the butt of the propeller blades causing the blades to rotate to a different angle.

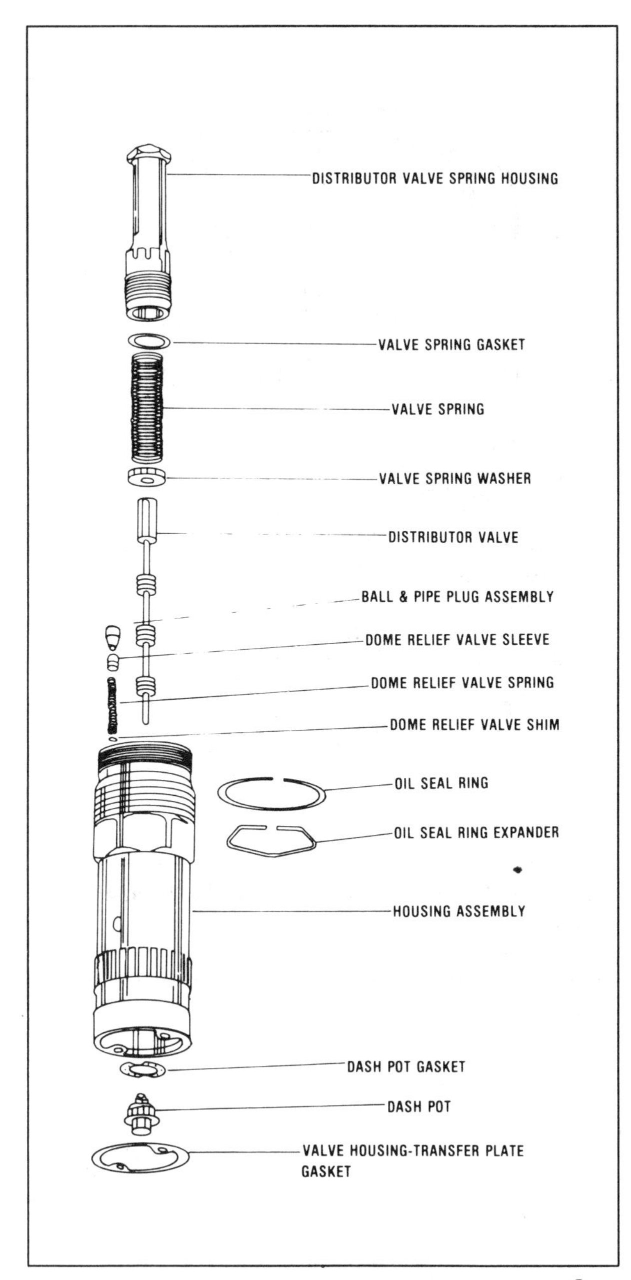

Figure 11-23. An exploded view of the Hydromatic® distributor valve.

The distributor valve is used to direct oil from the crankshaft to the inboard and outboard side of the piston and is shifted during unfeathering to reverse the oil passages to the piston.

2. Governor

The feathering Hydromatic® governor includes all of the basic governor components discussed for constant-speed governors. In addition, the Hydromatic® governor contains a high-pressure transfer valve which is used to block the governor constant-speed mechanism out of the propeller control system when the propeller is feathered or unfeathered.

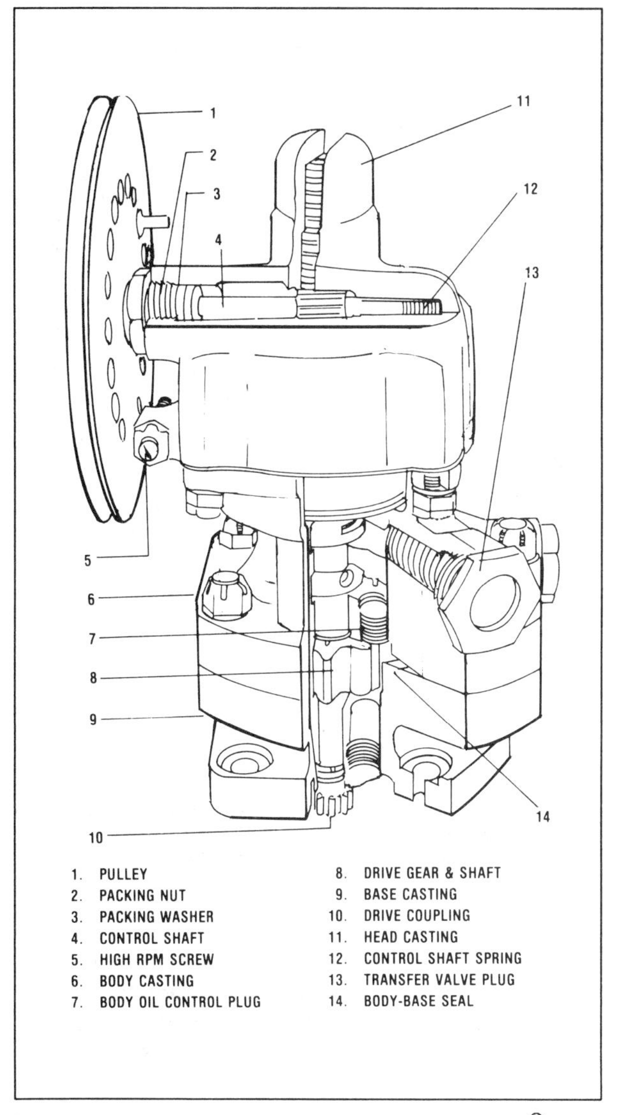

Figure 11-24. A cutaway view of the Hydromatic® feathering governor with a mechanical head.

A pressure cut-out switch is located on the side of the governor and is used to terminate the feathering operation automatically.

Hydromatic® governors may use mechanical heads (discussed in the Hamilton-Standard constant-speed section) or electric heads. Electric governor heads are controlled from the cockpit by toggle switches which are spring-loaded to the center-off position. When the toggle is pushed forward, the electric motor in the governor head rotates to reposition the governor speeder rack for a higher RPM setting. When the toggle switch is moved rearward, the governor head motor positions the speeder rack for a lower RPM.

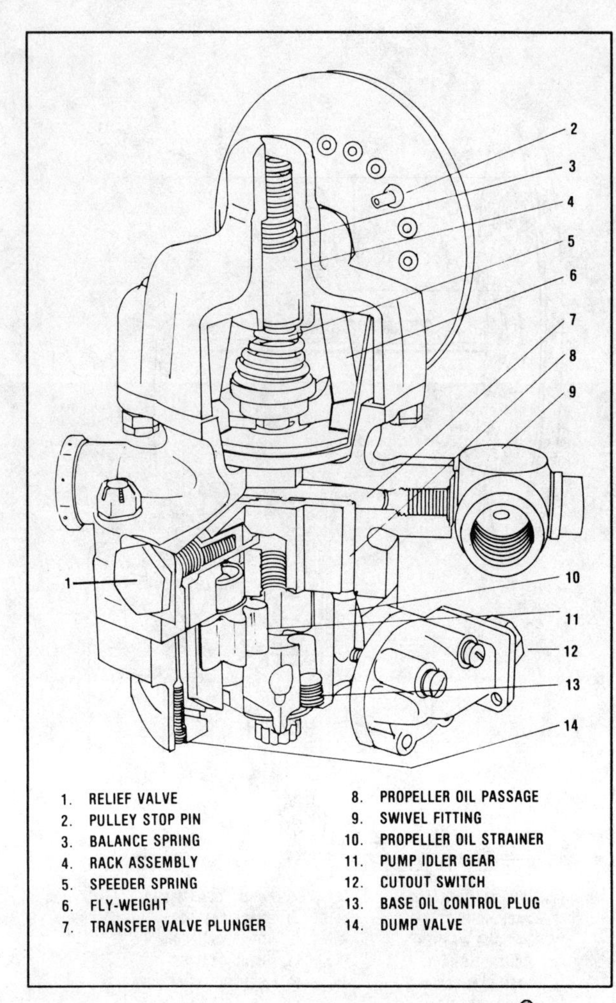

Figure 11-25. A cutaway view of a Hydromatic® feathering governor from a different angle showing transfer valve and pressure cutout switch.

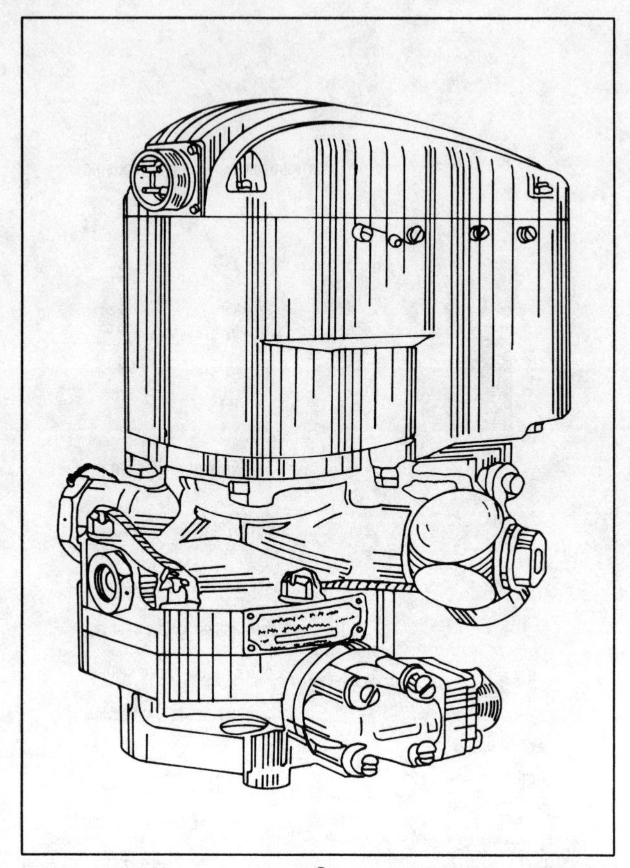

Figure 11-27. A Hydromatic® governor with an electric head.

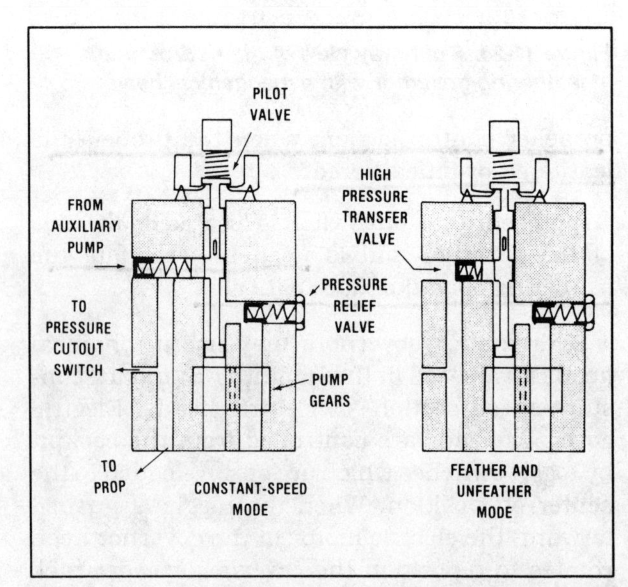

Figure 11-26. The high-pressure transfer valve blocks the governor constant-speed mechanism out of the system during feathering and unfeathering.

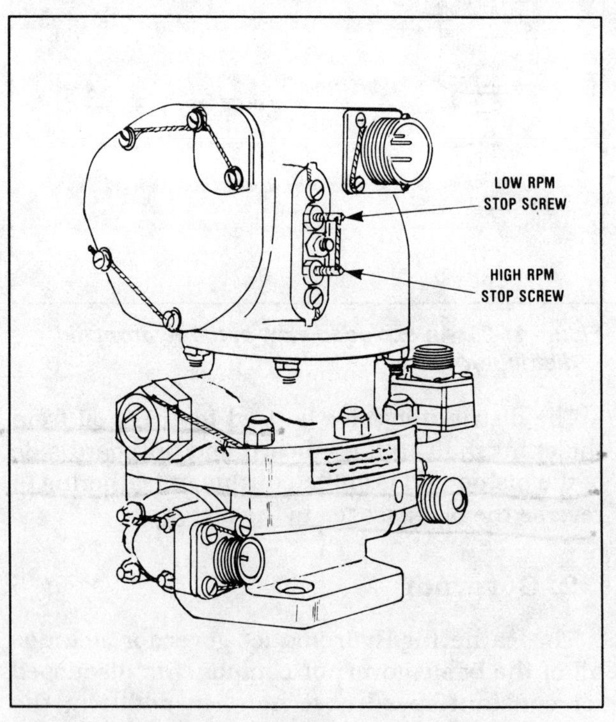

Figure 11-28. Adjust the RPM stop screws on the electric head of the governor and then safety them together.

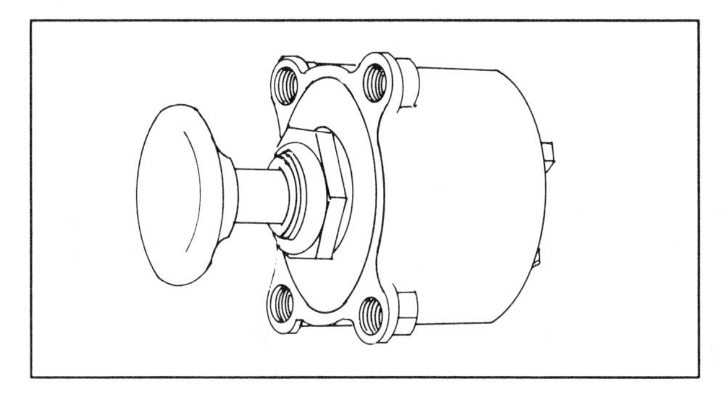

Figure 11-29. Cockpit feather button.

3. Feathering System Components

The cockpit control for the feathering system is simply a push button. This button is approximately $1^1/_4$ inches in diameter and is used to feather and unfeather the propeller. The feather button is usually located inside a shield to prevent accidental operation. The feathering button incorporates a holding coil to hold the button in electrically when the button is pushed.

A feathering relay is used to keep high currents out of the cockpit and reduce the length of large cables needed to get high current (200 amps or more) from the battery to the feathering (auxiliary) pump. The feathering relay is actuated when the feather button is pushed.

An electrically-operated feathering pump is used to supply oil under high pressure (about 600 psi) to the propeller when the feathering system is actuated. The pump draws oil from a standpipe in the engine oil supply tank.

4. System Operation

The Hydromatic® propeller uses governor oil pressure on one side of the propeller piston opposed by engine oil pressure on the other side of the piston aided by centrifugal twisting moments. Depending on the model of the propeller, governor oil pressure may be directed to the outboard side or inboard side of the piston.

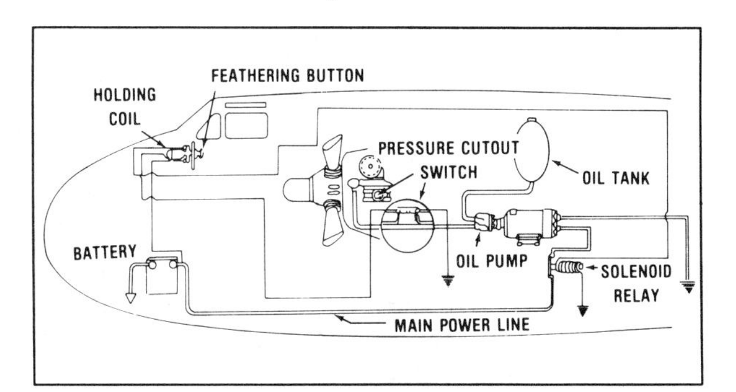

Figure 11-30. Feathering system configuration.

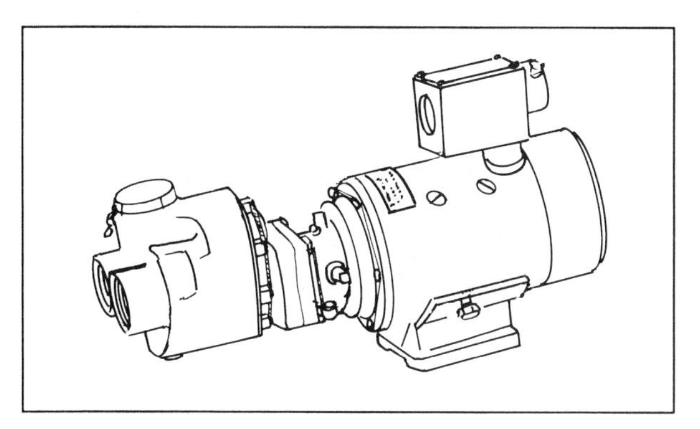

Figure 11-31. Auxiliary pump and motor used in the Hamilton-Standard feathering system.

For discussion purposes, consider that governor oil pressure is on the inboard side of the propeller piston and engine oil pressure is on the outboard side of the piston.

The Hydromatic® propeller does not use any springs or counterweights for operation. The fixed force is the engine oil pressure which is about 60 psi. The governor oil pressure (200 or 300 psi depending on the system) is controlled by the pilot valve during constant-speed operation.

When the system is in an overspeed condition, the pilot valve in the governor is raised and governor oil pressure is allowed to flow to the inboard side of the propeller piston via the crankshaft transfer bearing and the distributor valve. The increase in pressure on the inboard side of the piston causes the piston to move outboard. As the piston moves outboard, it rotates following the slot in the stationary cam and causes the rotating cam to rotate. As the rotating cam turns, the gears on the bottom of the cam mesh with the gears on the

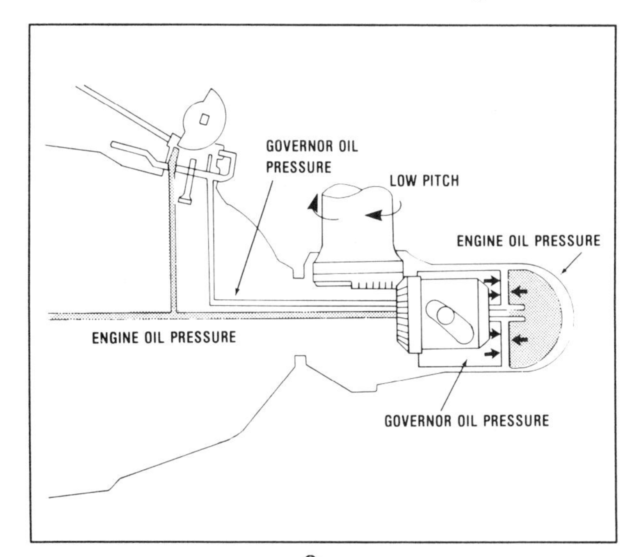

Figure 11-32. Hydromatic® propeller operating forces.

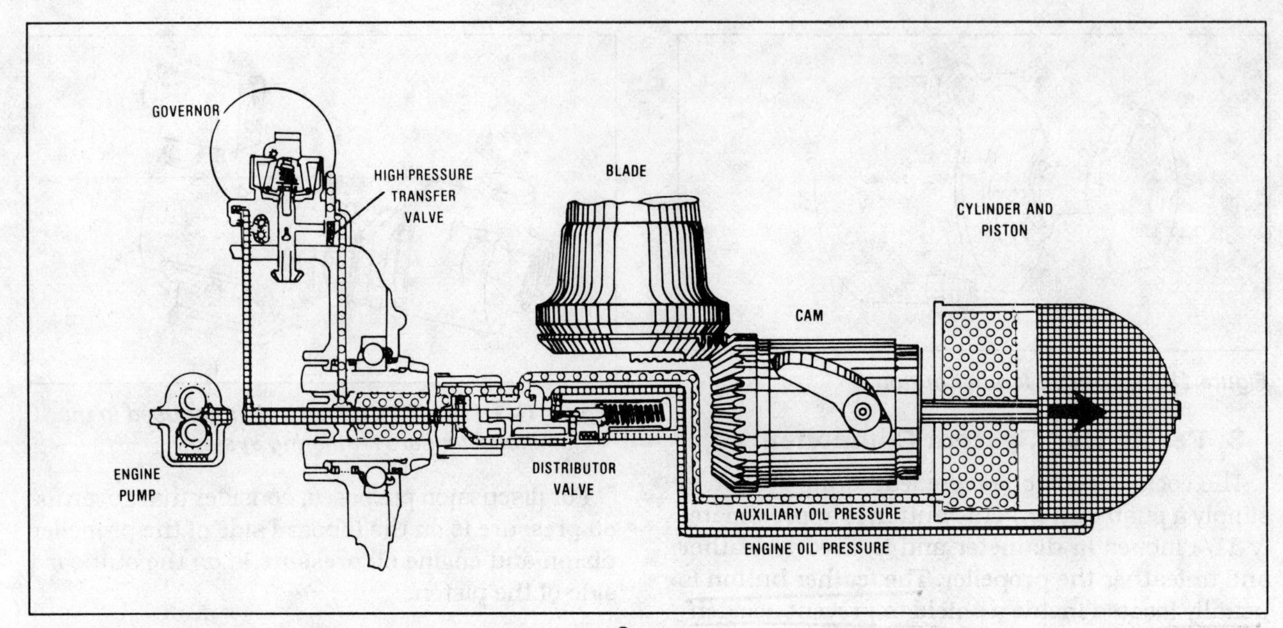

Figure 11-33. Feathering operations of the Hydromatic® propeller.

blade segment and cause the blade angle to increase. With this increase in blade angle, the system RPM slows down and the governor returns to the onspeed condition. The oil in the outboard side of the piston is forced back into the engine lubrication system where the pressure is maintained constant by the engine oil pressure relief valve.

When the system is underspeed, the pilot valve is lowered and the governor oil pressure in the inboard side of the piston is released. This causes engine oil pressure on the outboard side of the piston to force the piston inboard. As this piston moves inboard, the rotation created by the piston and the cams causes the blades to rotate to a lower blade angle allowing system RPM to increase to the onspeed condition.

To feather the propeller, the feather button in the cockpit is pushed. When this is done electrical contacts close and energize the holding coil which holds the feather button in. Another set of electrical contacts closes at the same time in the feather button and causes the feathering relay to close. The feathering relay completes the circuit from the battery to the feathering pump and the high pressure oil generated by the pump shifts the high-pressure transfer valve in the governor to block the governor out of the system. This high pressure oil is then directed to the inboard side of the piston and moves the blades toward the feather angle. When the rotating cam contacts the high pitch stop, the piston stops moving and the blades have reached the feather angle. Since the piston cannot move any further, the pressure in the system starts to build rapidly. This increasing pressure is sensed by the pressure cutout switch on the governor, which will break the circuit to the feather button holding coil at about 650 psi. This releases the feather relay and shuts off the feathering pump. All system pressures drop to zero and the blades of the propeller are held in feather by aerodynamic forces.

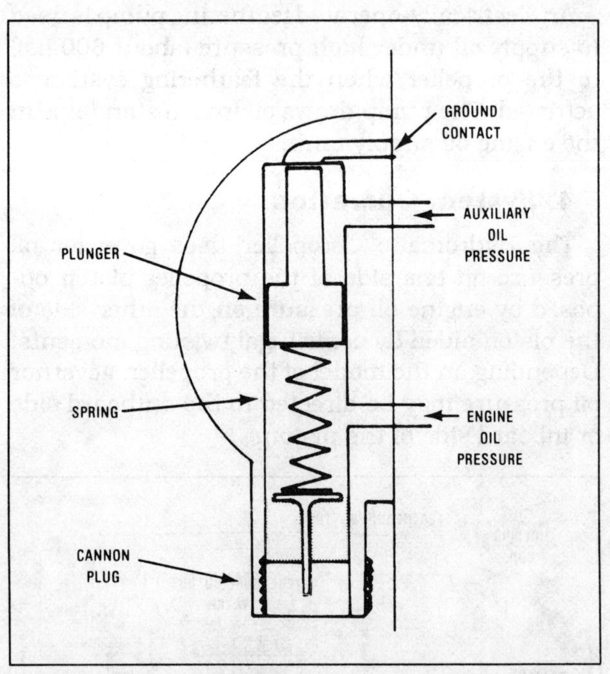

Figure 11-34. Cross section of a pressure cutout switch showing the spring and engine oil pressure on the plunger which opposes the force of the auxiliary oil pressure so that the feathering circuit does not break until the auxiliary pressure is very high.

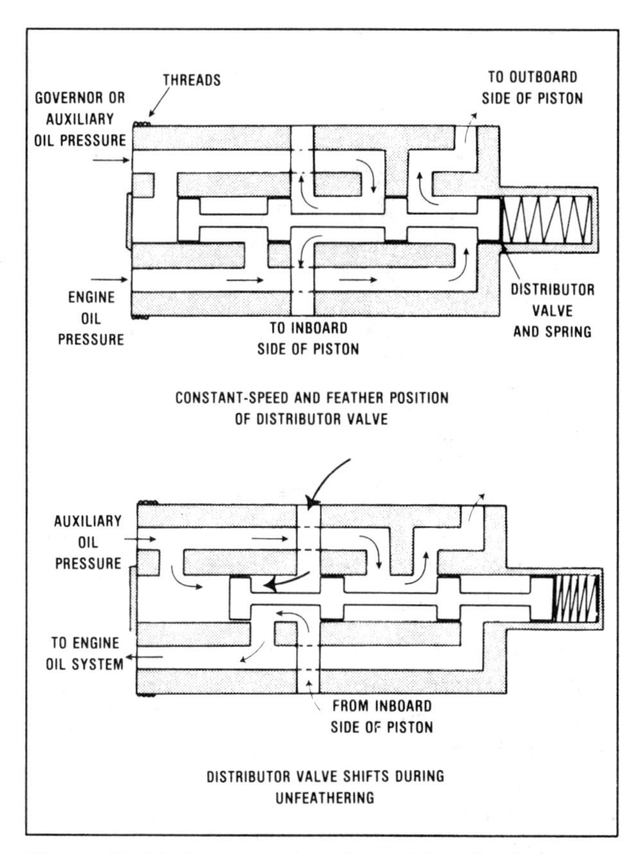

Figure 11-35. Cross section view of the distributor valve during propeller operations.

To unfeather the propeller, the feather button is pushed and held in to prevent the button from popping back out when the pressure cutout switch opens. The feathering pump starts building pressure above the setting of the pressure cutout switch, causing the distributor valve to shift and reverse the flow of oil to the piston. Feathering pump pressure is then directed to the outboard side of the piston and engine oil lines are open to the inboard side of the piston. The piston moves inboard and causes the blades to rotate to a lower blade angle through the action of the cams. With this lower blade angle the propeller starts to windmill, allowing the engine to be restarted. At this point the feather button should be released and the system will return to constant-speed operation. If the feather button is not released, the dome relief valve in the distributor valve will off-seat and release excess oil pressure (above 750 psi) from the outboard side of the piston to the inboard side of the piston after the rotating cam contacts the low blade angle stop.

5. Installation And Adjustments

The installation of a Hydromatic® propeller requires that the barrel assembly be installed first, following the basic procedure used to install a fixed-pitch propeller on a splined crankshaft. A hoist is required for most Hydromatic® propellers due to their size and weight. Special Hamilton-Standard tools are necessary to tighten the retaining nut. The standard checks for proper cone seating and spline wear are made when installing the barrel assembly. The retaining nut is then torqued, but not safetied at this time.

The distributor valve gasket and distributor valve are installed in the crankshaft. The distributor valve is carefully screwed into the internal threads on the crankshaft to avoid damaging the threads. The distributor valve is then torqued to the value specified for the particular installation.

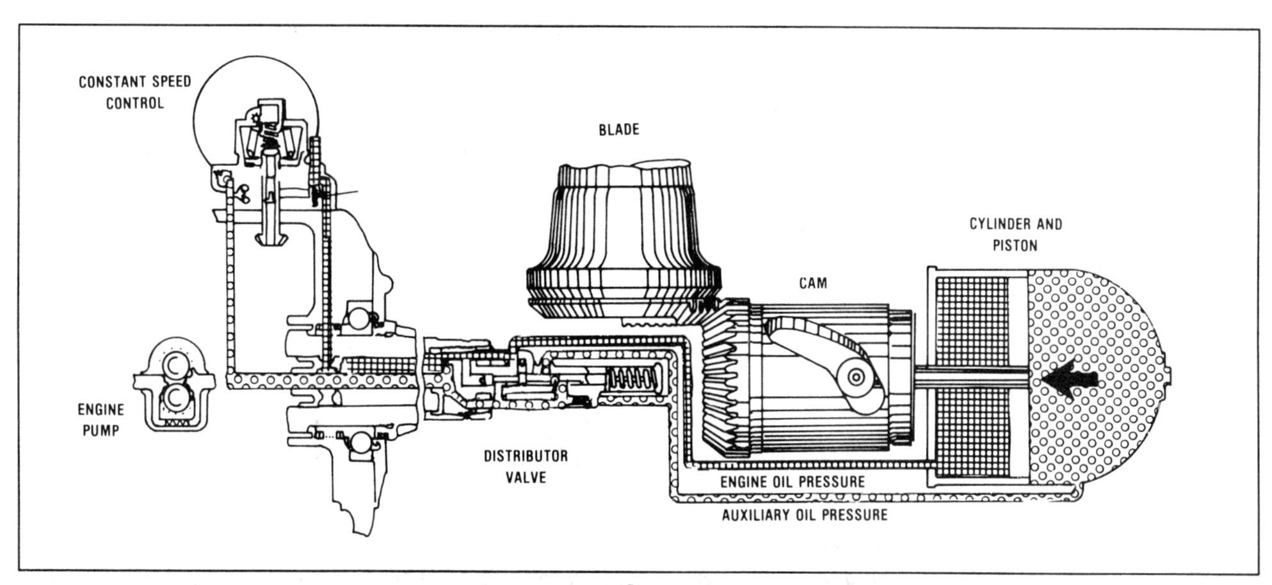

Figure 11-36. Unfeathering operation of the Hydromatic® propeller.

Figure 11-37. A hoist and sling are required to install most Hydromatic® propellers due to their size and weight.

Once the barrel assembly and distributor valve are installed on the crankshaft, a safety lock ring is inserted through a safety hole in the retaining nut and crankshaft, and into the safety slot of the distributor valve.

The dome assembly is now prepared for installation on the barrel assembly. The pitch stop rings must first be installed so that the propeller will have the proper blade angle range.

Refer to the proper aircraft or propeller service manual to determine the blade angles required

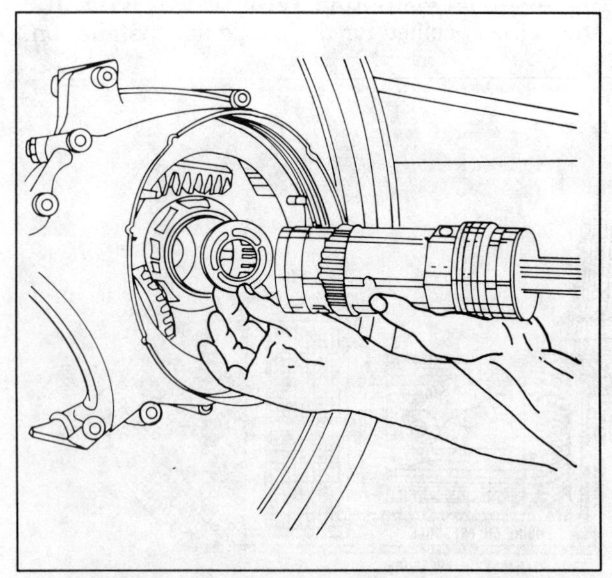

Figure 11-38. Position the distributor valve gasket inside the crankshaft and screw the distributor valve into the crankshaft.

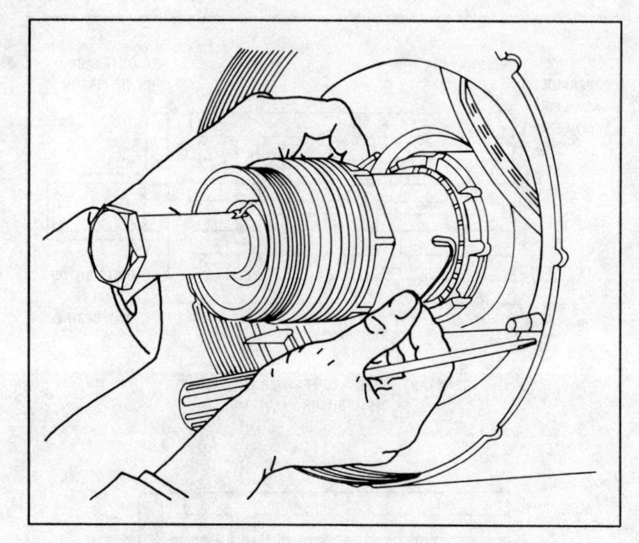

Figure 11-39. Installing the barrel and distributor valve locking ring.

and the location of the reference tooth on the stop rings. The reference tooth on the stop rings is used as the index which is placed next to the blade angle scale on the rotating or stationary cam. (The location varies with propeller models). The reference tooth may be either to the right (+) or to the left (–) of the zero reference tooth and may or may not be marked by a scribed line or arrow. The zero reference tooth is the tooth which is in the center of the stop ring lug on most models (see Figure 11-40.)

For discussion purposes, presume that the low pitch stop ring is indexed at –23, the high pitch stop ring is indexed at +7, and the blade angles are 18 and 95 degrees. From the zero reference tooth on the low pitch stop ring, count 23 teeth to the left and mark the tooth with a red lead or white lead pencil. Place the stop ring on the base of the dome assembly so that the reference tooth lines up with the 18-degree mark on the cam scale and install the ring fully into the dome base. Count seven teeth to the right of the zero reference tooth on the high pitch stop ring

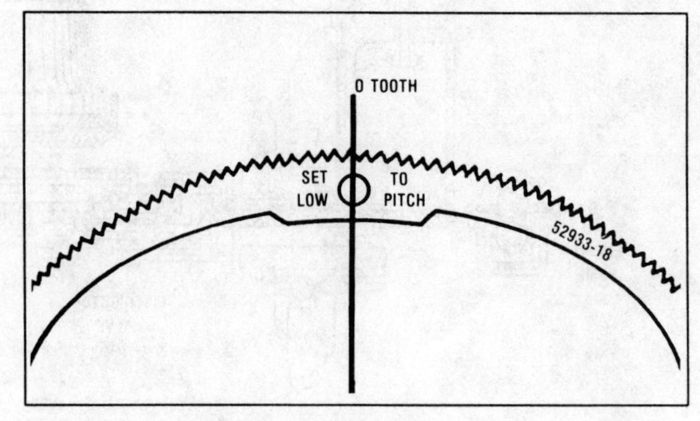

Figure 11-40. The zero reference tooth is located in the center of the stop ring lug.

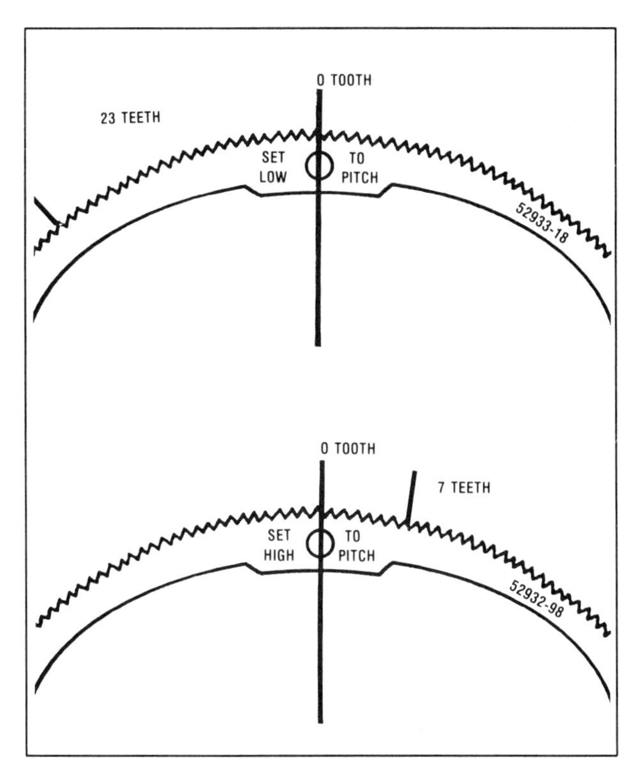

Figure 11-41. Locate the reference tooth on the stop ring by counting from the zero reference tooth.

and mark this tooth. The high pitch stop ring is positioned so that the reference tooth aligns with the 95-degree mark on the cam scale and the ring is then pressed into the dome base on top of the low pitch stop ring. The piston should be in a mid-range position when installing the stop rings.

The rotating cam is turned until the lugs on the high pitch stop ring contact the dome stop lugs. This sets the dome in the feather position.

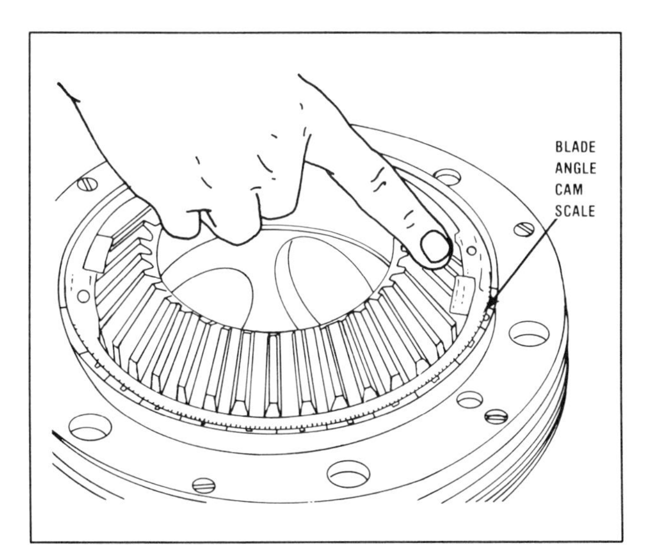

Figure 11-42. Install the stop rings in the base of the propeller dome.

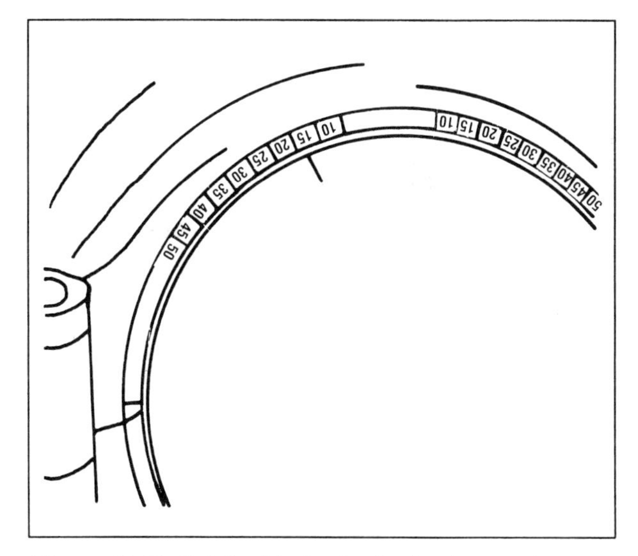

Figure 11-43. Set the blades to the feather angle using the scale on the blades at the hub.

The blades in the barrel are rotated until the scale on the blade shank indicates that the blades are at the feather angle. The dome shim is installed on the dome shelf in the barrel and the base gasket is placed on the bottom of the dome. The dome is installed on the barrel following the procedure for the particular propeller model, then it is torqued and safetied. The dome plug is installed, torqued, and safetied.

The propeller track and the high and low blade angles of each blade are then checked.

The governor is installed and rigged in the same manner as for the Hamilton-Standard constant-speed governor. The only additional steps required are to connect the cannon plug to the pressure cutout switch and attach the oil line from the feathering pump.

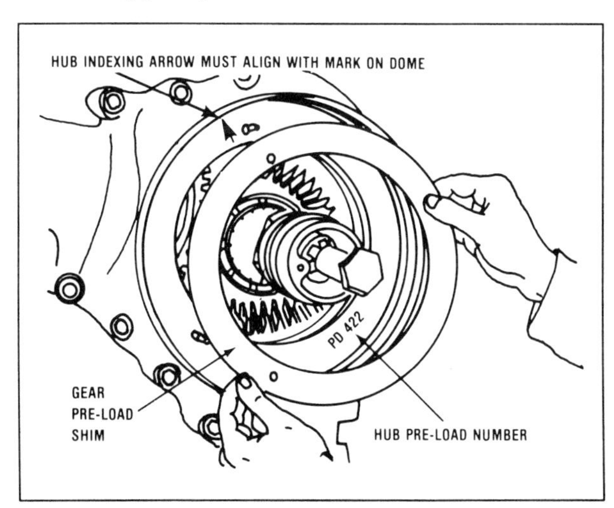

Figure 11-44. Install the dome shim on the barrel shelf before installing the dome assembly.

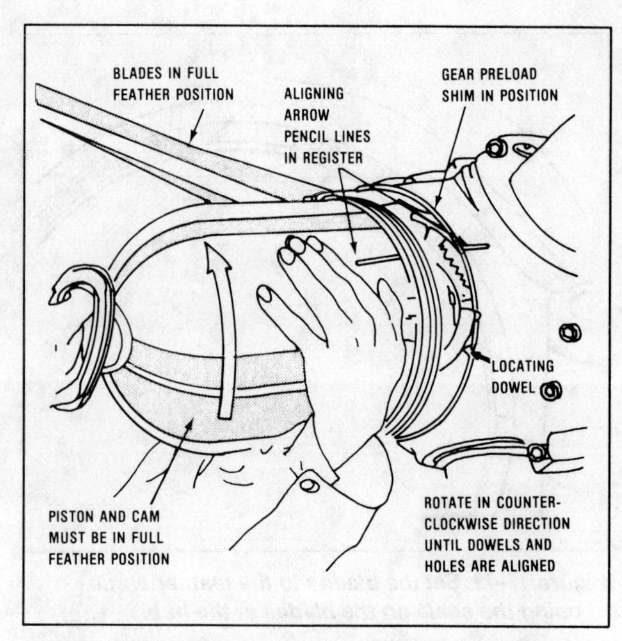

Figure 11-45. Install the dome assembly on the barrel using a lifting handle on the dome.

If an electric head is used on the governor, governor rigging is simplified. Since the cannon plug for the head needs only to be connected and the high and low RPM limit switches set.

The feathering pump, feather button, and other system components are installed and adjusted according to the particular aircraft service manual.

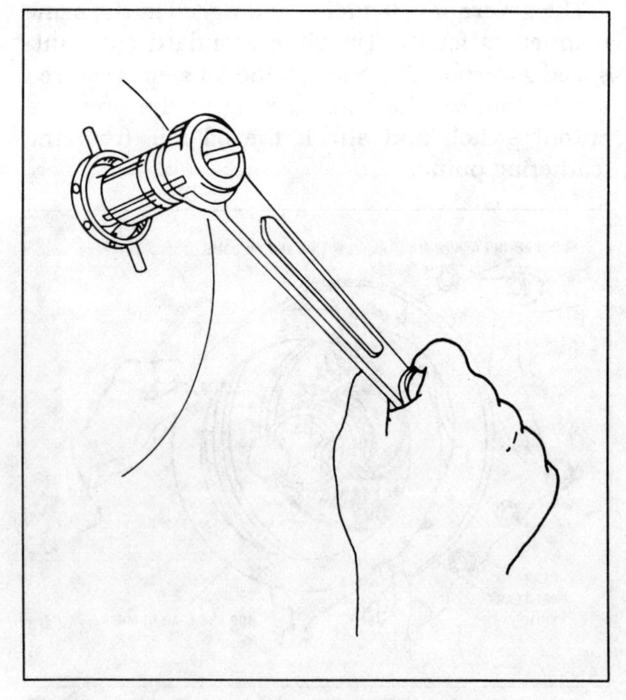

Figure 11-46. Remove the dome lifting handle and install the dome plug and required seals.

6. Inspection, Maintenance And Repair

The propeller and governor are inspected and repaired in accordance with the procedures discussed in previous chapters of this book. The inspection primarily involved assuring proper operation, checking for oil leaks, and inspecting the external oil lines for signs of deterioration and abrasion.

Oil leaks in the propeller are normally caused by a faulty gasket or a loose nut or bolt. If oil covers all of the propeller, the dome plug is leaking. If oil appears on the barrel immediately behind the dome, the dome gasket is leaking or the dome nut is loose. These defective gaskets can be replaced in the field and the nuts should be torqued and safetied.

If oil is coming from the blade shank area or from between the barrel halves, the hub bolts may be loose or the gaskets may be defective. If no irregularities are found, the bolts may be retorqued. The gaskets must be replaced by an overhaul facility.

The propeller is lubricated by engine operating oil and does not need to be lubricated during maintenance.

7. Troubleshooting

Troubleshooting procedures and solutions discussed for other systems are generally applicable to the feathering Hydromatic® system.

If the propeller fails to respond to the cockpit propeller control lever, but can be feathered and unfeathered, the cause is most likely a failure of the governor or governor control system. To locate this problem, check the control system first. If it is functioning properly, replace the governor.

If the propeller fails to feather, check the system for electrical faults or open wiring to the electrical components. Other causes may be a defective feathering pump or a stuck high-pressure transfer valve in the governor. Replace the defective component, desludge the governor, or replace the governor, as appropriate.

If the propeller fails to unfeather after feathering normally, the distributor valve is not shifting. Replace the distributor valve.

If the propeller feathers and immediately unfeathers, the problem may be a shorted line from the holding coil to the pressure cutout switch or a defective pressure cutout switch. The same effect will result if the feather button is shorted internally. The feather button or pressure cutout switch may be replaced as necessary.

Sluggish movement of the propeller may be the result of a buildup of sludge in the propeller dome or a worn out piston-to-dome seal in the dome. The sludge can be removed by removing the dome and cleaning with a solvent. A worn piston-to-dome seal requires replacement by an overhaul facility.

Erratic or jerky operation of the propeller is an indication of the wrong preload shim being used between the dome and barrel assemblies. The dome should be removed and the proper shim should be installed.

QUESTIONS:

1. *What forces are used to feather the McCauley propeller?*

2. *What is the purpose of the accumulator in the light aircraft feathering systems?*

3. *What are the two shapes of accumulators?*

4. *If an accumulator is not used in a light aircraft feathering system, what pilot technique may help to unfeather the propeller in flight?*

5. *What is the approximate air pressure in an accumulator?*

6. *What forces may be used to feather the Hartzell Compact propeller?*

7. *What cockpit procedure is followed to feather the propeller on a light aircraft?*

8. *What are the three major assemblies used in a Hydromatic® propeller?*

9. *Which Hydromatic® propeller assembly contains the propeller pitch-changing mechanism?*

10. *What is the purpose of the high-pressure transfer valve on the Hydromatic® governor?*

11. *Which Hydromatic® governor component automatically terminates the feathering operation?*

12. *What force opposes governor oil pressure in the Hydromatic® propeller?*

13. *During which Hydromatic® propeller operation does the distributor valve shift?*

14. *What is the position of the Hydromatic® propeller piston when installing the pitch stop rings?*

15. *What is the position of the Hydromatic® propeller blades when the dome assembly is installed on the barrel assembly?*

Chapter XII
Reversing Propeller Systems

A *reversing propeller* is a constant-speed feathering propeller with the additional capability of producing a reverse thrust.

Reversing propeller systems are used on most modern multi-engine turboprop aircraft such as the Cessna Conquest, Beechcraft King Air, Piper Cheyenne, and on large transport aircraft such as the Douglas DC-7 and the Lockheed Constellation. Some small single-engine seaplanes and floatplanes use reversing propellers for improved water maneuverability.

Reversing propellers have the advantages of decreasing the length of the landing roll, reducing brake wear, and increasing ground maneuverability. Some aircraft can use the reversing system to back up the aircraft on the ground, while other designs use the system only to brake the aircraft on the landing roll.

The main disadvantages of the reversing system are the reduced engine cooling available for piston-driven aircraft, and increased blade damage from stones, sand, etc., when the propellers are in reverse.

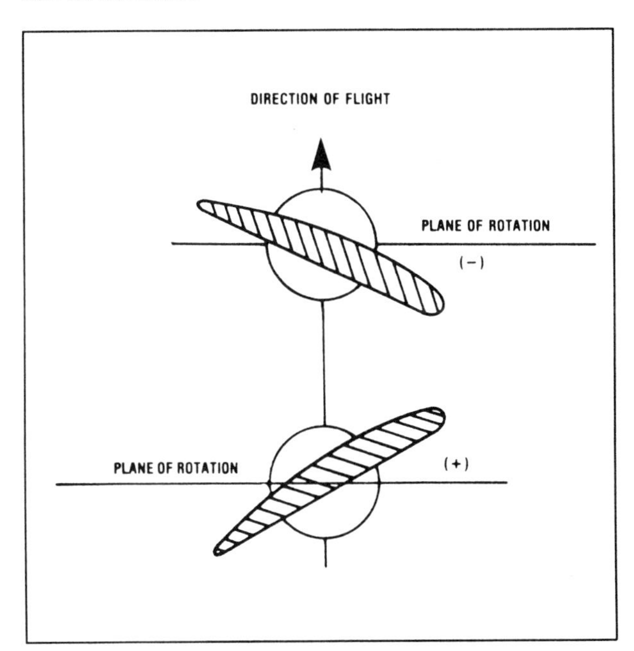

Figure 12-1. The reverse angles are negative angles compared to positive angles in the constant-speed range and the feather angle.

When a propeller goes into reverse, the blades rotate below the low blade angle and into a negative angle of about −15 degrees. This forces air forward to provide a negative thrust. When the system goes into reverse, the engine does not rotate in the direction opposite to normal rotation (as some people think), the blades just force air forward rather than rearward.

The reversing operation is controlled in the cockpit by the throttles and is initiated by moving the throttles aft of the idle position. This reverses the blades and the engine RPM and/or the blade angle is varied by moving the throttles within the reverse range to control the amount of reverse thrust. The farther aft the throttles are moved, the greater the reverse thrust.

Propellers cannot normally be reversed in flight and the aircraft often must be below a specified airspeed on the landing roll before the reverse mechanism is engaged. Most systems require that the aircraft weight be on the landing gear before the throttles can be moved into the reverse range. This is controlled by a squat switch on the landing gear strut.

This chapter will cover the Hartzell reversing propellers used with the Garrett AiResearch TPE-331, the Pratt & Whitney PT6 engines, and the Hamilton-Standard reversing Hydromatic® propeller system used on transport aircraft.

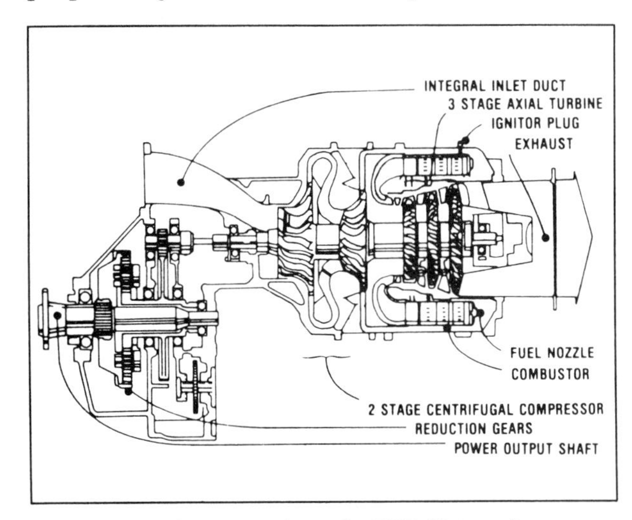

Figure 12-2. Cutaway view of a TPE-331 engine.

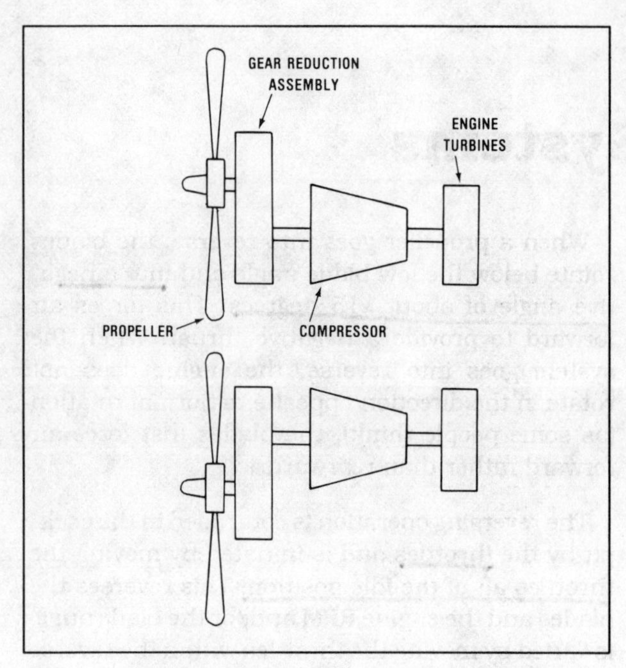

Figure 12-3. Basic components of the engine and the two installation positions.

A. Hartzell Reversing Propeller System On The Garrett AiResearch TPE-331 Engine

The Hartzell propeller on the TPE-331 is used on aircraft such as the Mitsubishi MU-2, Short Skyvan, and the Cessna Conquest.

The TPE-331 engine is a fixed turbine engine of over 600 horsepower at an engine RPM of about 40,000. A gear reduction assembly on the front of the engine couples the engine driveshaft to the propeller driveshaft and reduces the engine RPM to about 2,200 RPM at the propeller drive shaft. (These values vary somewhat with different models of the engine.) The engine may be installed with the propeller driveshaft above or below the engine centerline as shown in Figure 12-3.

1. Propeller

The propeller commonly used on the TPE-331 is a three- or four-bladed Hartzell Steel Hub

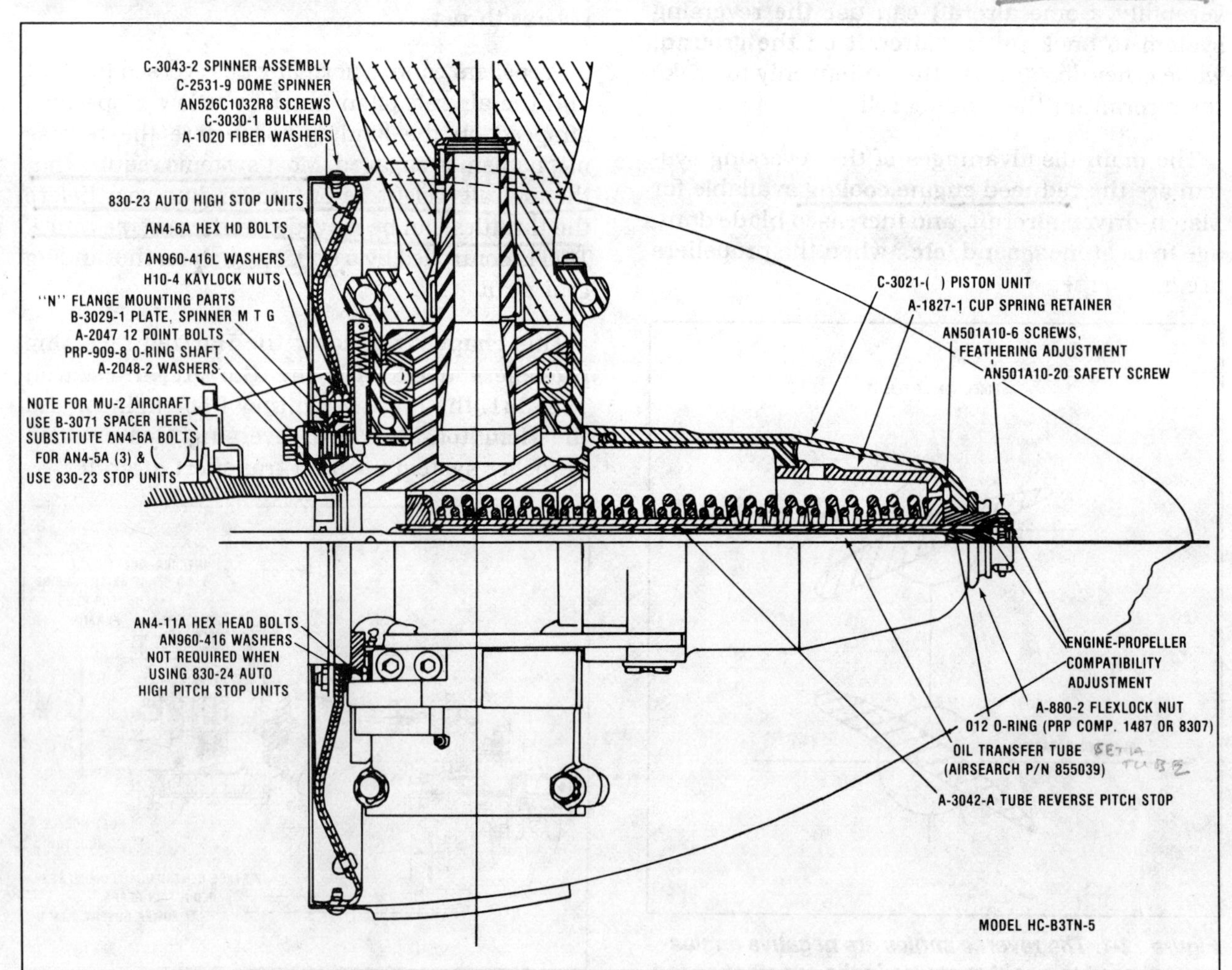

Figure 12-4. Cutaway view of the Hartzell propeller used on the TPE-331 engine.

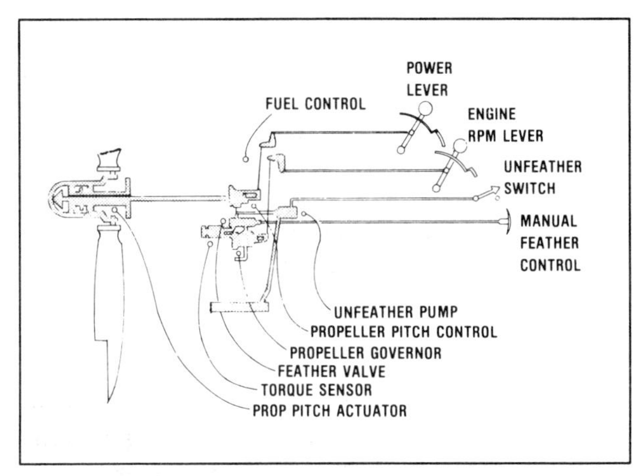

Figure 12-5. The location of the system components on the gear reduction assembly of the TPE-331 engine.

reversing propeller. The propeller is spring-loaded and counterweighted to the feather position and uses engine oil pressurized by the governor to decrease the blade angle. It is flange-mounted on the driveshaft and locks in a *flat* angle of about two degrees when the engine is shut down on the ground. This prevents excessive stress on the engine starter system when starting the engine.

The propeller is constructed similar to the feathering Steel Hub designs. The principal additional component is the Beta tube which passes through the center of the propeller and serves as an oil passage and follow-up device during propeller operation.

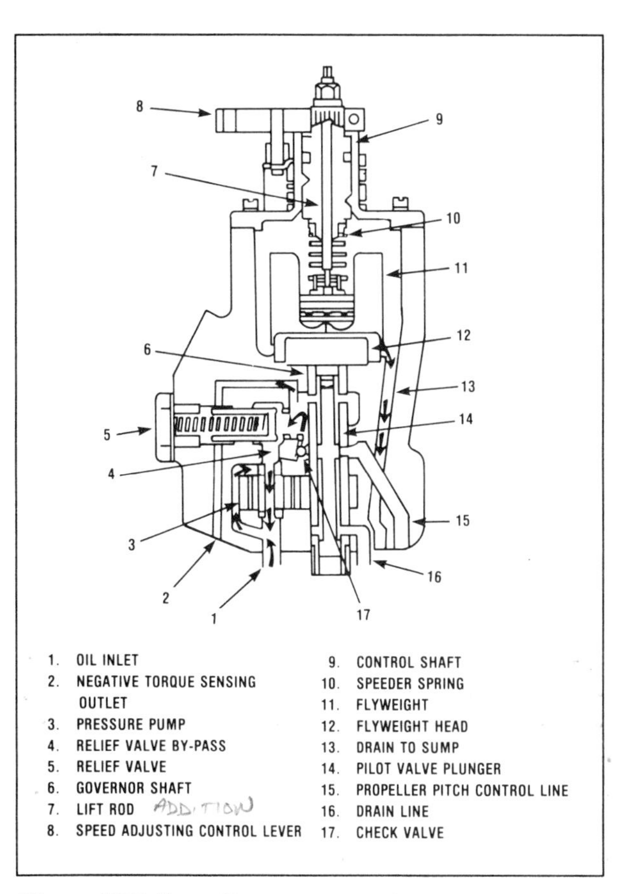

Figure 12-7. Propeller governor cutaway.

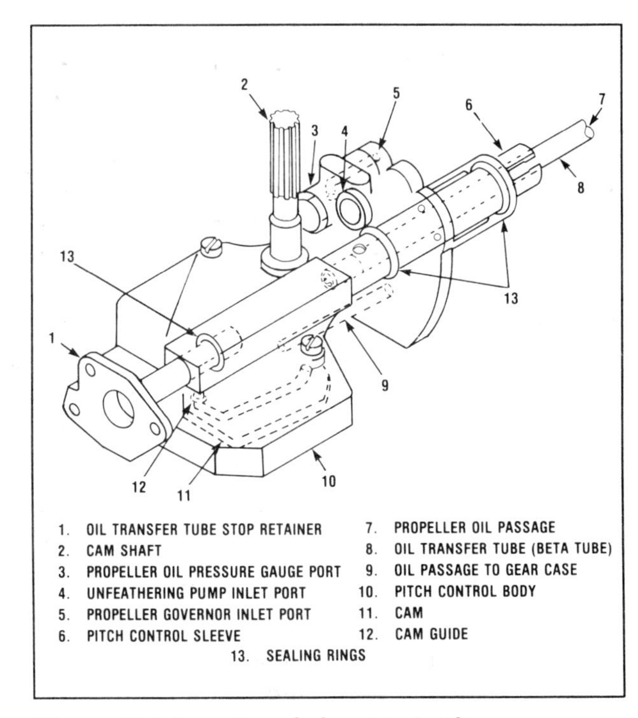

Figure 12-6. Propeller pitch control unit.

The same designation system is used for the reversing Hartzell propellers as is used for Hartzell feathering and constant-speed propellers.

2. System Components

The propeller pitch control is mounted on the rear of the gear reduction assembly in line with the propeller driveshaft and is connected to the propeller through the Beta tube. The propeller pitch control is operated by the cockpit power lever and is used to direct oil to and from the propeller to change blade angles during ground operations. During flight operations the propeller pitch control serves only as an oil passage between the propeller and the propeller governor.

The propeller governor is mounted on the gear reduction assembly and operates the same way as other constant-speed governors to control system RPM in flight (through a range of about 2,000 to 2,200 RPM). Below 2,000 RPM the propeller governor is Underspeed and serves only to provide oil pressure to the propeller pitch control.

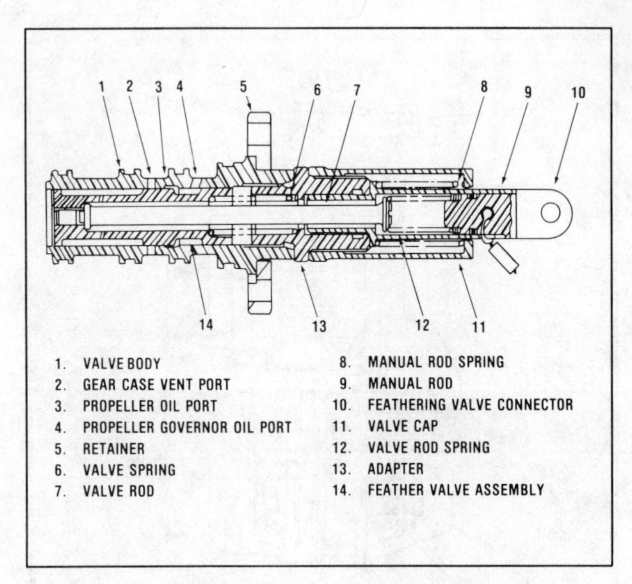

Figure 12-8. Cross section of the feathering valve.

The underspeed governor, which is part of the fuel control unit, is used to control system RPM when the propeller is being driven at less than 2,000 RPM. The underspeed governor is operated by the condition lever and controls fuel flow to the engine to maintain a selected RPM below the propeller governor range.

A feathering valve is operated automatically by a torque sensor in the engine or manually from the cockpit to release all oil pressure from the propeller allowing the springs and counterweights to feather the propeller.

An electric unfeathering pump is used to supply oil pressure to unfeather the propeller.

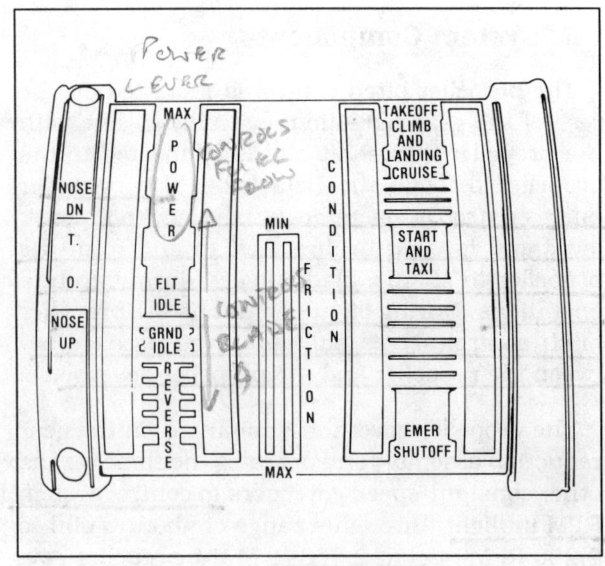

Figure 12-9. Cockpit powerplant quadrant for the TPE-331 powered Cessna Conquest.

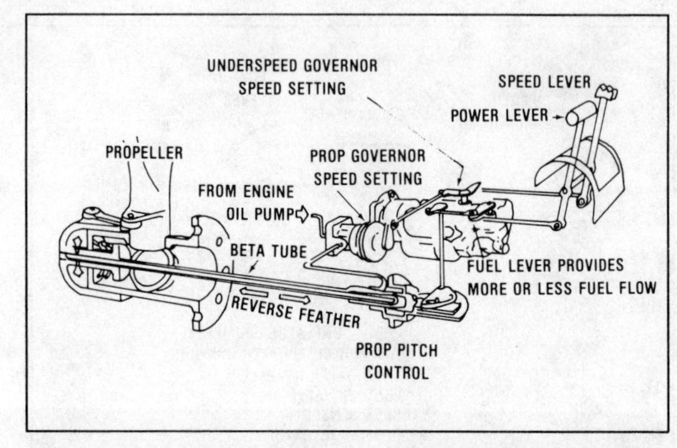

Figure 12-10. Control system for the TPE-331 system showing the cockpit control and the accessories.

3. Cockpit Controls

The cockpit controls for the TPE-331 turboprop installation include a power lever which controls the horsepower output of the engine, a condition lever which controls system RPM, a feather handle, and an unfeathering switch.

The power lever is similar to the reciprocating engine throttle in that it controls system horsepower. During ground operation the power lever directly controls the propeller blade angle by positioning the propeller pitch control. During flight operations, the power lever directly controls fuel flow to the engine through the engine fuel control unit.

The condition lever is similar to the propeller control lever in a reciprocating engine system in that it controls system RPM. During ground operation the condition lever adjusts the underspeed governor on the fuel control unit to vary the fuel flow and maintain a fixed RPM as the blade angle is changed by the power lever. During flight operations the condition lever sets the propeller governor to maintain system RPM by varying the blade angles when the engine power is changed with the power lever or when flight operations change.

Many aircraft use a feather handle connected to the feathering valve on the engine. Other aircraft connect the feathering valve to the condition lever so that full aft movement of the lever will cause the propeller to feather. When the feathering valve is moved by the cockpit control, oil is released from the propeller and the propeller feathers.

An unfeathering switch is used to control the electric unfeathering pump to unfeather the propeller.

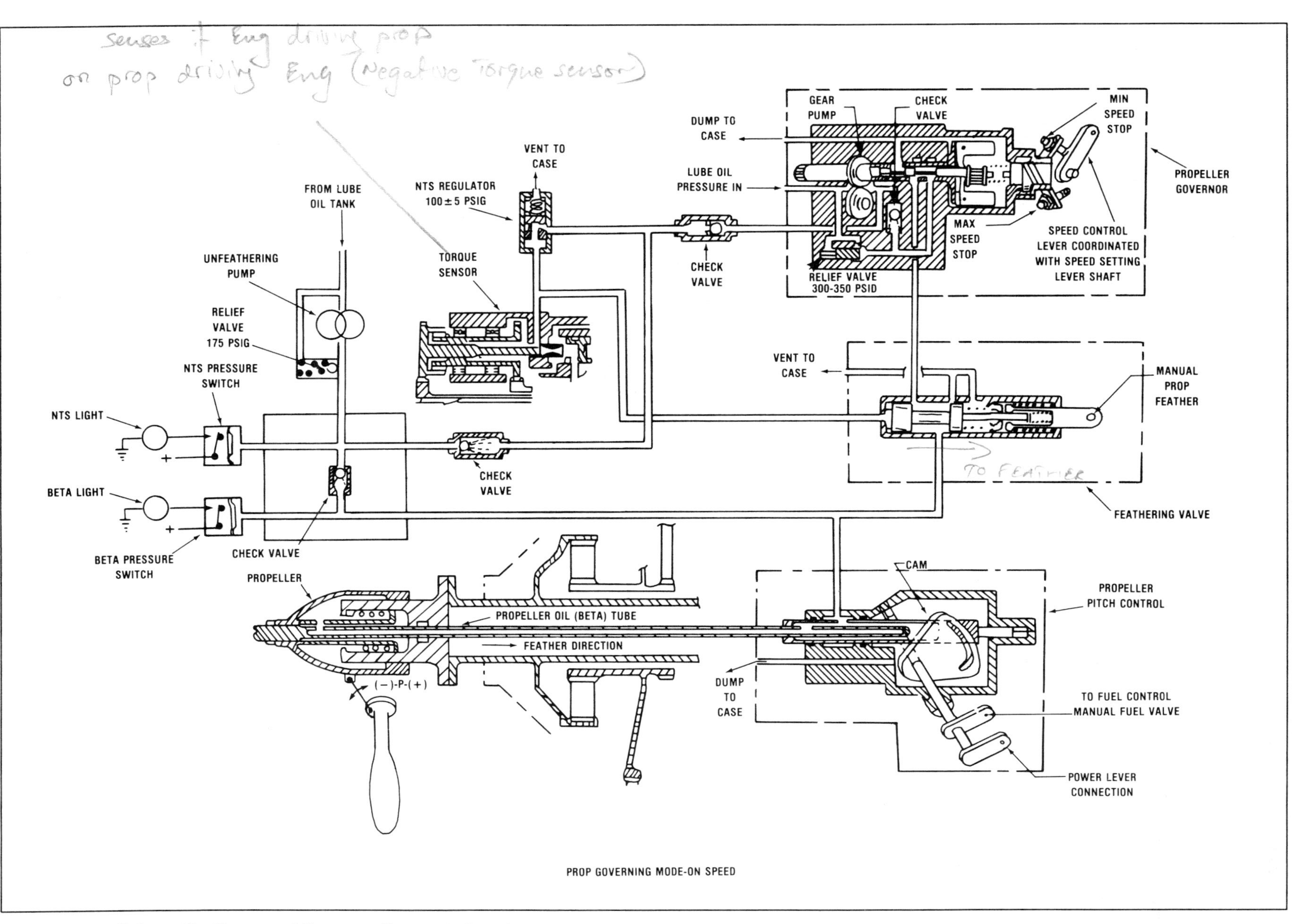

Figure 12-11. Schematic diagram of the propeller control system.

4. System Operation

The two basic operating modes of the TPE-331 system are the Beta Mode, meaning any ground operation including start, taxi, and reverse operation, and the Alpha Mode, meaning any flight operation from takeoff to landing. Typically, Beta Mode includes operation from 65% to 95% and Alpha Mode includes operation from 95% to 100% of system rated RPM.

When the engine is started, the power lever is set at the ground idle position and the condition lever is in the start position. When the engine starts the propeller latches retract and the propeller moves to a zero degree blade angle as the propeller pitch control is positioned by the power lever over the Beta tube. The Beta tube is attached to the propeller piston and moves forward with the piston as the propeller moves to the low blade angle. The propeller blade angle stops changing when the Beta tube moves forward to the neutral position in the propeller pitch control.

The condition lever is used to set the desired RPM through the underspeed governor during ground operations, and the power lever is used to vary the blade angle to cause the aircraft to move forward or rearward. If the power lever is moved forward, the propeller pitch control moves rearward so that the oil ports on the end of the Beta tube are open to a drain line to the engine sump and the oil in the propeller is forced out by the springs and counterweights. As the blade angle increases, the propeller piston moves inward, moving the Beta tube inward until it returns to the neutral position. This causes a proportional response of the propeller to the power lever movement.

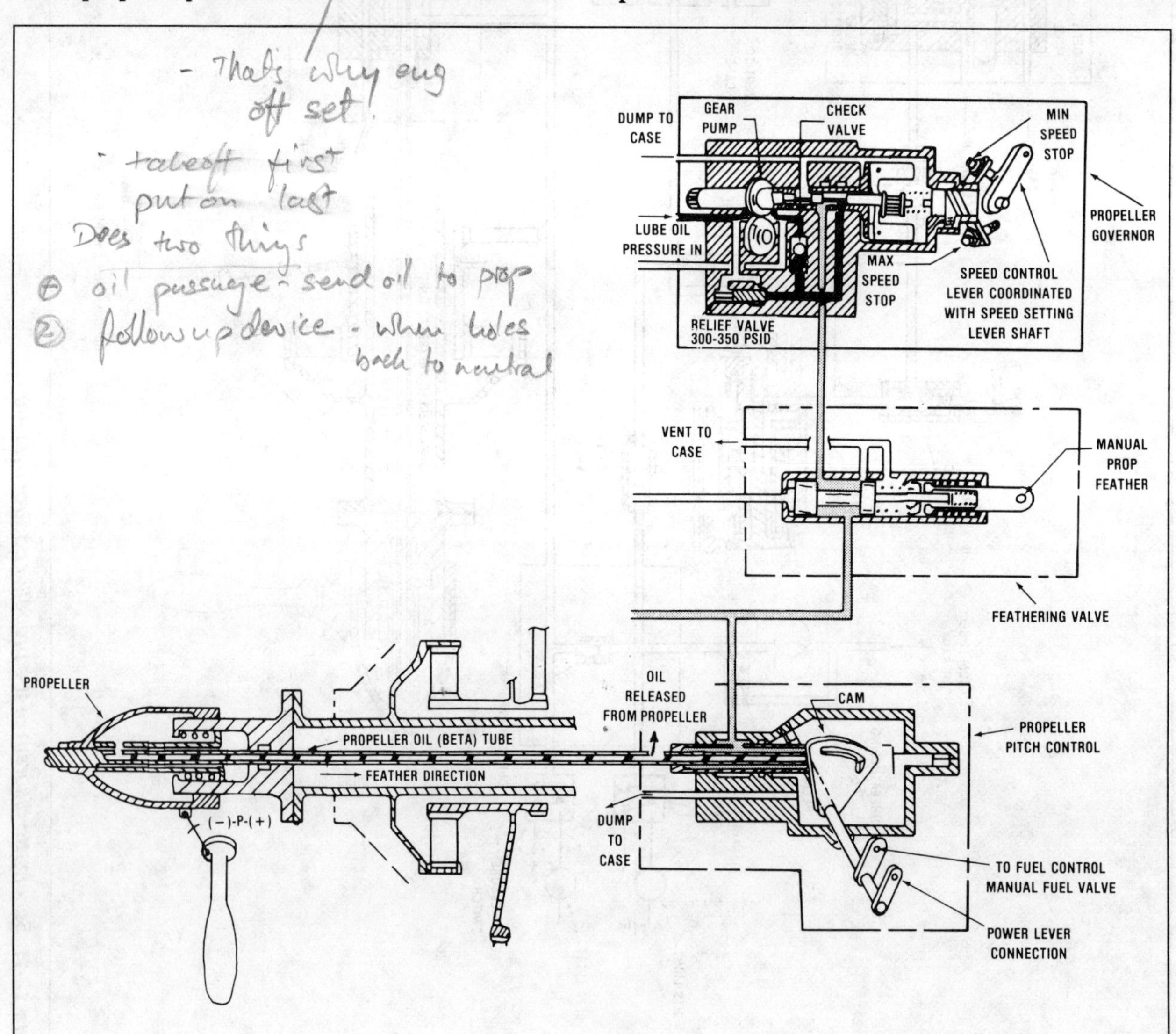

Figure 12-12. System components positioned to increase propeller blade angle in the Beta Mode.

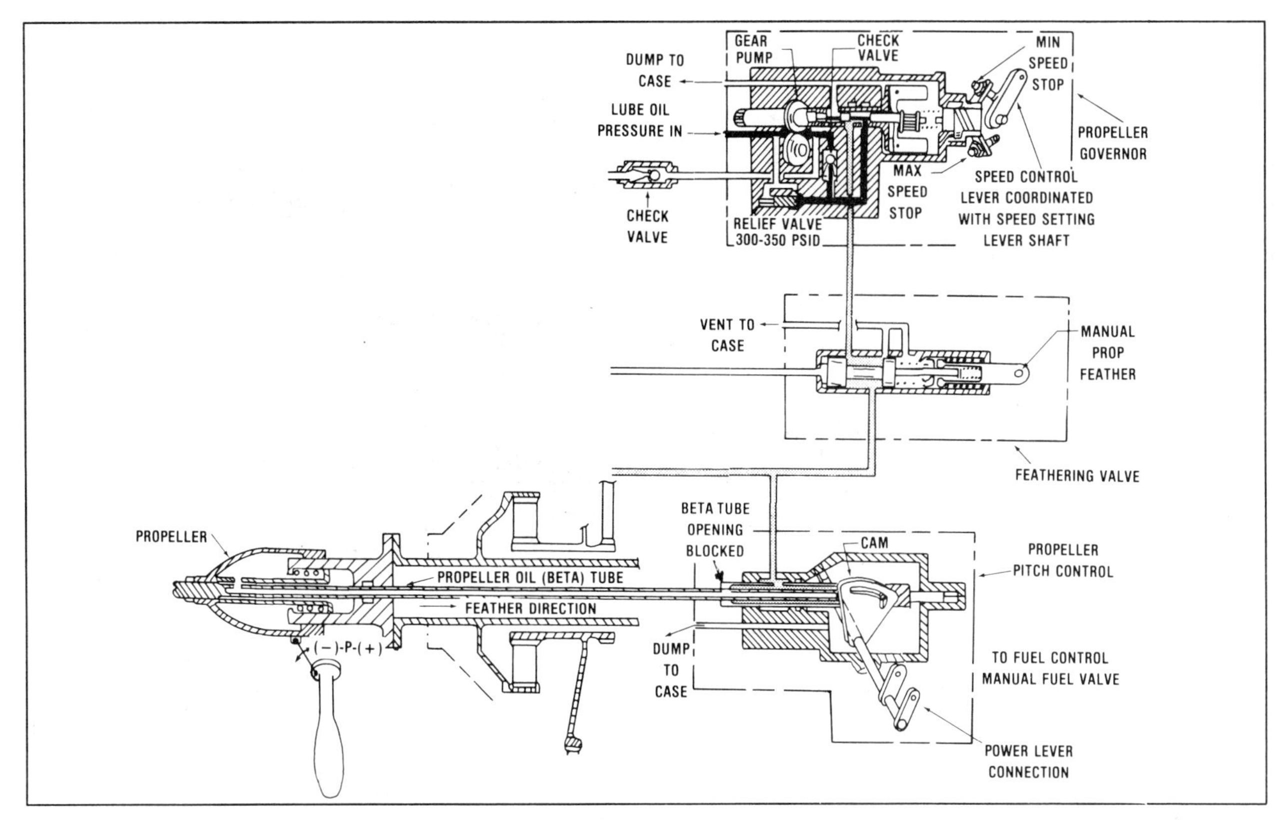

Figure 12-13. The Beta Tube stops propeller blade angle change in the Beta Mode by moving to the neutral position in the propeller pitch control unit.

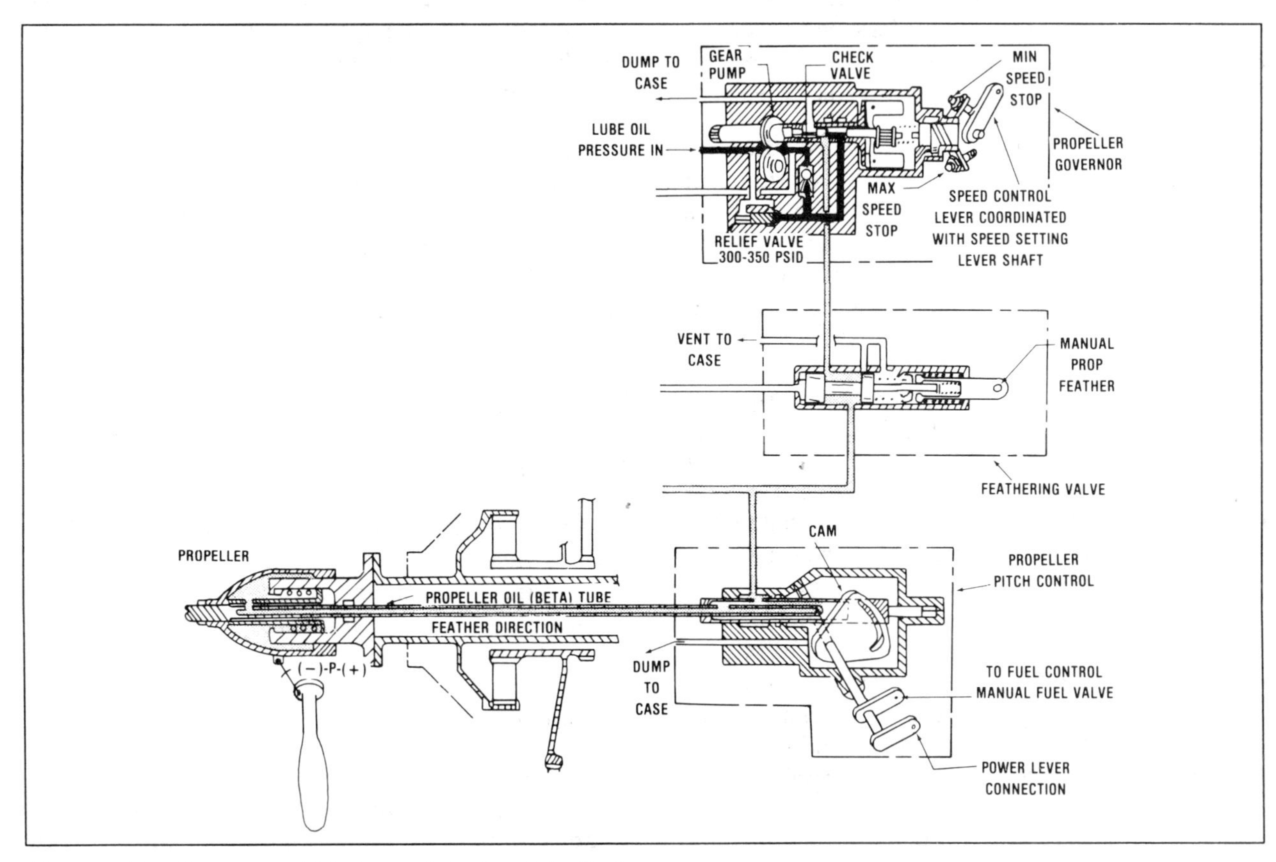

Figure 12-14. Decrease of blade angle in the Beta Mode.

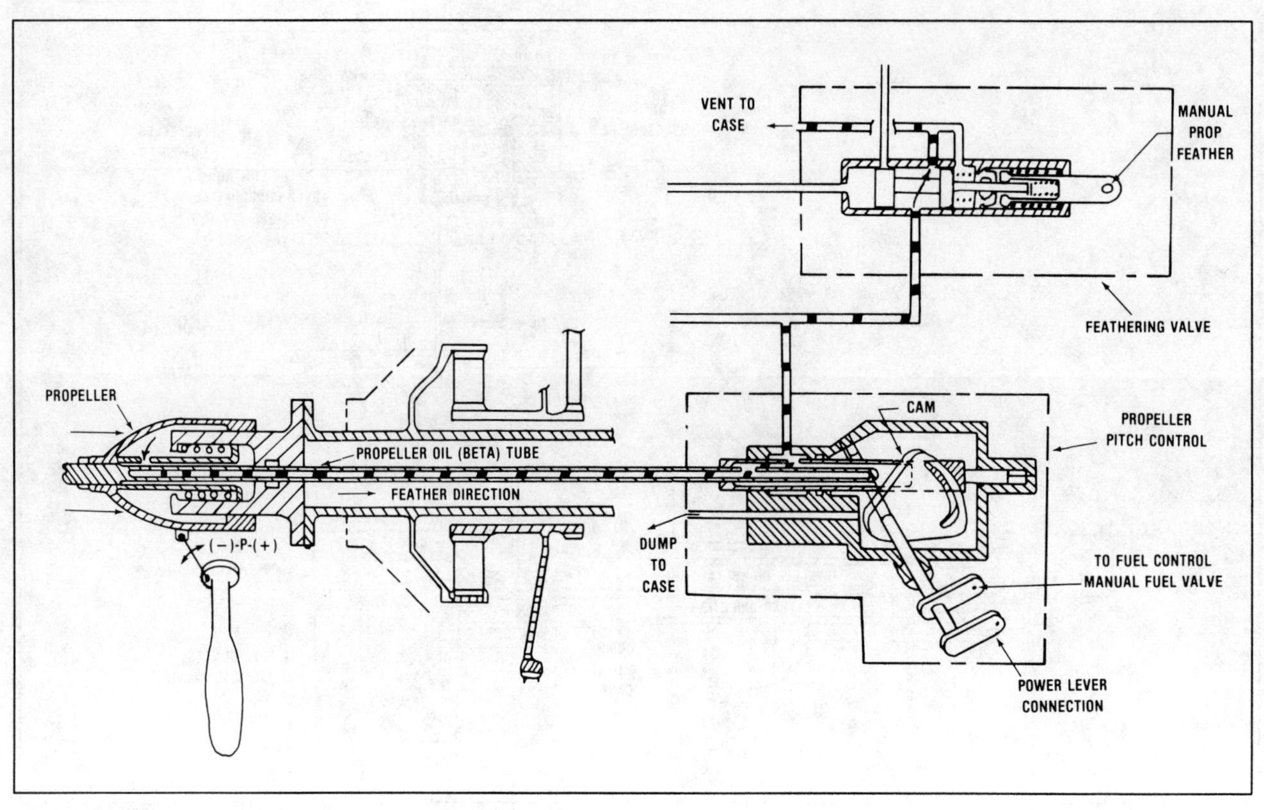

Figure 12-15. Feathering of the propeller by the cockpit control.

With the increase in blade angle the engine will start to slow down, but the underspeed governor, which is set by the condition lever, will increase fuel flow to the engine to maintain the selected RPM.

If the power lever is moved rearward, the propeller pitch control moves forward over the Beta tube and governor oil pressure flows out to the propeller piston and causes a decrease in blade angle. As the piston moves outward, the Beta tube moves with it and will return to the neutral position as the blade angle changes. With this lower blade angle the engine RPM will tend to increase, but the underspeed governor will reduce the fuel flow to maintain the selected RPM.

In the Alpha Mode of operation (flight operations), the condition lever is moved to a high RPM setting (95% to 100%) and the power lever is moved to the flight idle position. When this is done, the underspeed governor is opened fully and no longer affects system operation. RPM control is now accomplished through the propeller governor. When the power lever is moved to flight idle, the propeller pitch control moves forward so that the Beta tube is fully in the propeller pitch control and no longer functions to adjust blade angle. The power lever then controls fuel flow through the fuel control unit.

With a fixed power lever setting in the Alpha Mode, the propeller governor is adjusted by the condition lever to set the system RPM in the same as for any constant-speed system.

With a fixed condition lever setting in the Alpha Mode, the power lever adjusts the fuel control unit to control the amount of fuel delivered to the engine. If the power lever is moved forward, fuel flow will increase and the propeller blade angle will be increased by the propeller governor to absorb the increase in engine power and maintain the set RPM. If the power lever is moved aft, fuel flow will decrease and the propeller blade angle will decrease by the action of the propeller governor to maintain the selected RPM.

Whenever it is desired to feather the propeller, the feather handle is pulled or the condition lever is moved full aft, depending on the aircraft design. This action shifts the feathering valve, located on the side of the gear reduction assembly, and releases the oil pressure from the propeller, returning the oil to the engine sump.

The springs and counterweights on the propeller force the oil out of the propeller and the blades go to the feather angle. The feather valve may be operated hydraulically by the engine

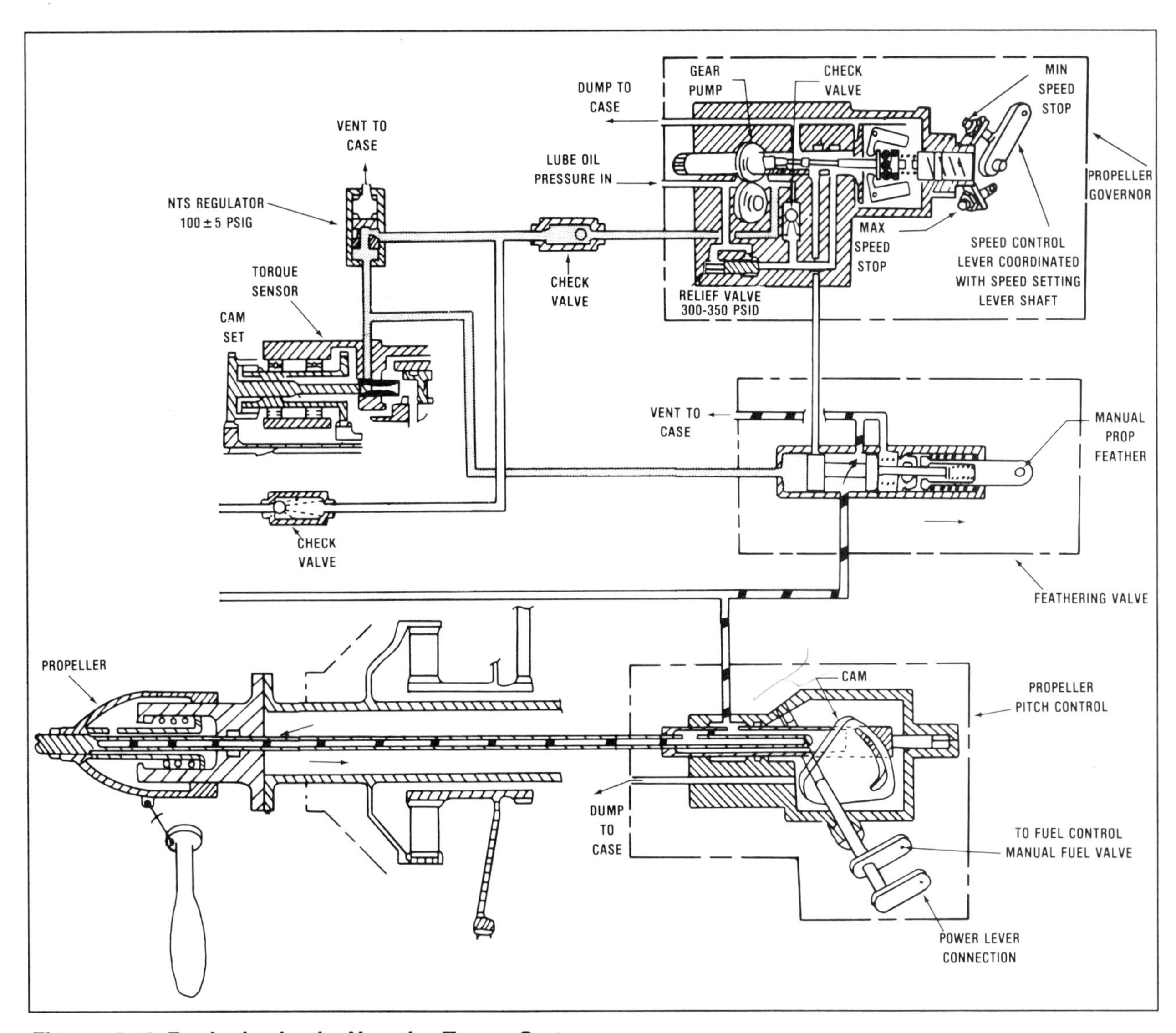

Figure 12-16. Feathering by the Negative Torque System.

Negative Torque Sensing (NTS) system, sensing a loss of positive torque, directing oil pressure to the feathering valve and shifting it to the feather position. This system works automatically and requires no action by the pilot (Figure 12-16).

To unfeather the propeller, the electric unfeathering pump is turned on by a toggle switch in the cockpit and oil pressure is directed to the propeller to reduce the blade angle. This will cause the propeller to start windmilling in flight and an air start can be accomplished. On the ground the propeller can be unfeathered in the same manner before starting the engine.

5. Installation And Adjustment

The propeller is installed following the basic procedure used for the installation of flanged-shaft propellers. The Beta tube is installed through the propeller piston after the propeller is installed and is bolted to the forward part of the piston.

The adjustment of the propeller low pitch latch is performed by loosening the latch plates and shifting their position to give a two- or three-degree blade angle (depending on the blade model) at the 30-inch station.

To adjust the reverse blade angle, rotate the blades in the blade clamps as for a feathering and constant-speed Steel Hub propeller. This will cause the feather and low pitch stop angle to change also. The reverse angle can also be adjusted by an overhaul facility by changing the length of the tube reverse pitch stop inside the feathering spring assembly.

The feather angle can be adjusted by rotating the blades in the blade clamp. This will also change the low pitch stop and reverse angle to change. The feather angle can also be adjusted by removing the propeller cylinder and adjusting the screw on the front of the spring assembly. The screw will increase the feather angle 1.2 degrees for each turn into the spring assembly.

The propeller governor, propeller pitch control, feather valve, and fuel control unit are mounted on the engine gear reduction assembly in accordance with the engine service manual.

The interconnection between the condition lever, the underspeed governor, and the propeller governor is rigged and adjusted according to the manual pertaining to the particular aircraft and engine model used. The same holds for the interconnection between the power lever, the fuel control unit, and the propeller pitch control.

6. Inspection, Maintenance And Repair

Inspect and repair the propeller following the basic procedures set forth for other versions of the Hartzell Steel Hub propeller. Take care when removing and installing the Beta tube to prevent damage to the tube surface. The Beta tube is trued for roundness and is machined to close tolerances.

Inspect the propeller control units for leaks, security, and damage. Check the linkages between these units for freedom of movement, security, and damage. Replace defective seals, adjust rigging, and secure all nuts and bolts as appropriate for the installation. Use the engine or aircraft maintenance manuals for specific inspections which vary with different aircraft models.

7. Troubleshooting

Basic troubleshooting procedures as have been previously covered apply to the Hartzell reversing system. If the proper propeller response is not occurring, check the system for proper rigging if there is no obvious defect.

In the Beta Mode, if the RPM is not constant, investigate the underspeed governor on the fuel control unit. If the blade angle does not respond properly to power lever movement, check the propeller pitch control.

In the Alpha Mode if the RPM is not constant check the propeller governor. If power does not change smoothly, check the fuel control unit.

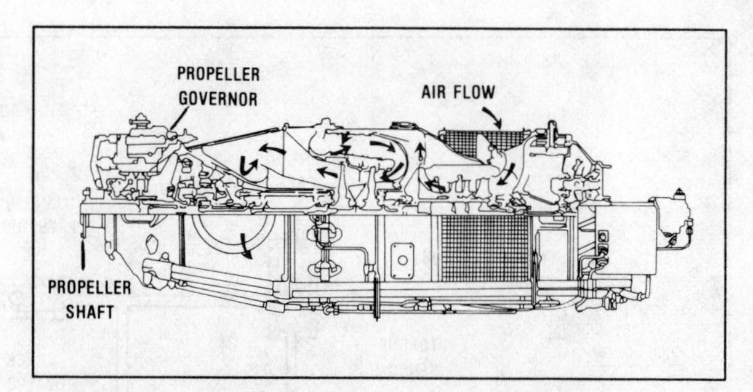

Figure 12-17. A PT6 engine.

B. Hartzell Reversing Propeller System On The Pratt & Whitney PT6 Engine

The Hartzell propeller on the PT6 engine is used on the Piper Cheyenne, DeHavilland Twin Otter, and most models of the Beechcraft King Air series.

The PT6 engine is a free turbine design of over 600 HP at 38,000 RPM. A gear reduction mechanism couples the engine power turbine to the propeller driveshaft with the propeller rotating at 2,200 RPM at 100% RPM. The engine is a free turbine design, meaning that the power turbine is not mechanically connected to the engine compressor, but is air coupled. The hot gases generated by the engine flow over the power turbine wheel and cause the power turbine and the propeller to rotate.

Another turbine section is mechanically linked to the compressor section and is used to drive the compressor section. It is possible during engine start for the compressor and its turbine to be rotating while the propeller and the power turbine do not move or move at a lower RPM. The power turbine will eventually reach the speed of the compressor, but the starter motor is not under a

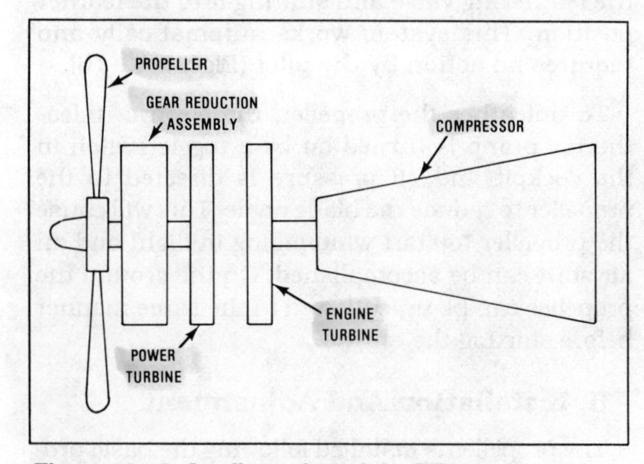

Figure 12-18. Configuration of the PT6 engine and propeller.

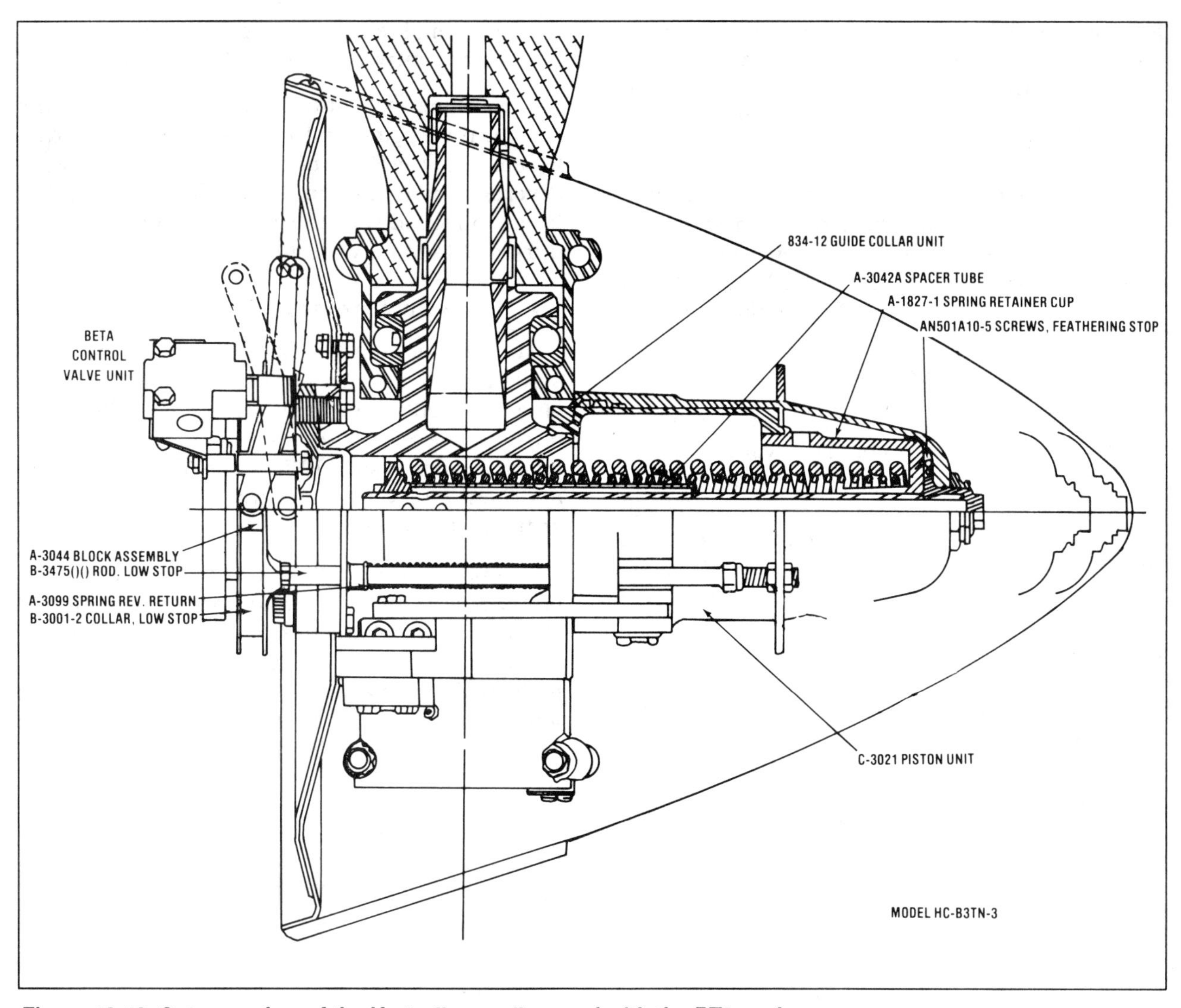

Figure 12-19. Cutaway view of the Hartzell propeller used with the PT6 engine.

load from the propeller and power turbine during engine start. For this reason the propeller can be shut down in feather and does not need a low blade angle latch mechanism for engine starting.

1. Propeller

The propeller commonly used with the PT6 is a three-, four-, or five-bladed Hartzell Steel Hub reversing propeller. The propeller is flange-mounted on the engine, and is spring-loaded and counterweighted to the feather position with oil pressure being used to decrease the blade angle. A slip ring mechanism on the rear of the propeller serves as a follow-up mechanism in giving proportional propeller response to control inputs in the beta mode.

2. Governor

The propeller governor used with the PT6 is basically the same as other governors discussed

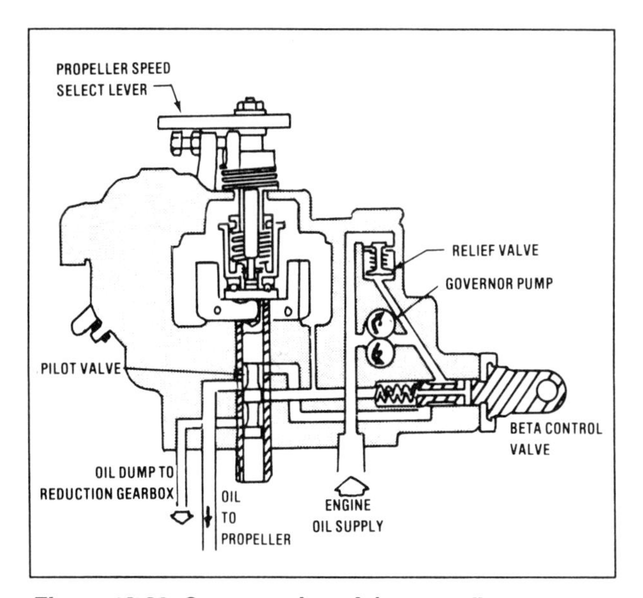

Figure 12-20. Cross section of the propeller governor for a PT6 installation.

for constant-speed operation, using a speeder spring and flyweights to control a pilot valve which directs oil flow to and from the propeller. A lift rod is incorporated in the governor to allow feathering of the propeller.

For Beta Mode operation the governor contains a *Beta control valve* operated by the power lever linkage and directs oil pressure generated by the governor boost pump to the propeller or relieves oil from the propeller to change the blade angle.

3. System Components

A propeller *overspeed* governor is mounted on the gear reduction assembly and will release oil from the propeller whenever the propeller RPM exceeds 100%. The release of oil pressure will result in a higher blade angle and a reduction in RPM. The overspeed governor is adjusted on the ground when the unit is installed and cannot be adjusted in flight. There are no cockpit controls to this governor.

A *power turbine governor* is installed on the gear reduction assembly as a safety device in case the other propeller governing devices should fail. When the power turbine RPM reaches about 105%, the power turbine governor will reduce fuel flow to the engine to prevent excessive engine RPM. The power turbine governor is not controllable from the cockpit.

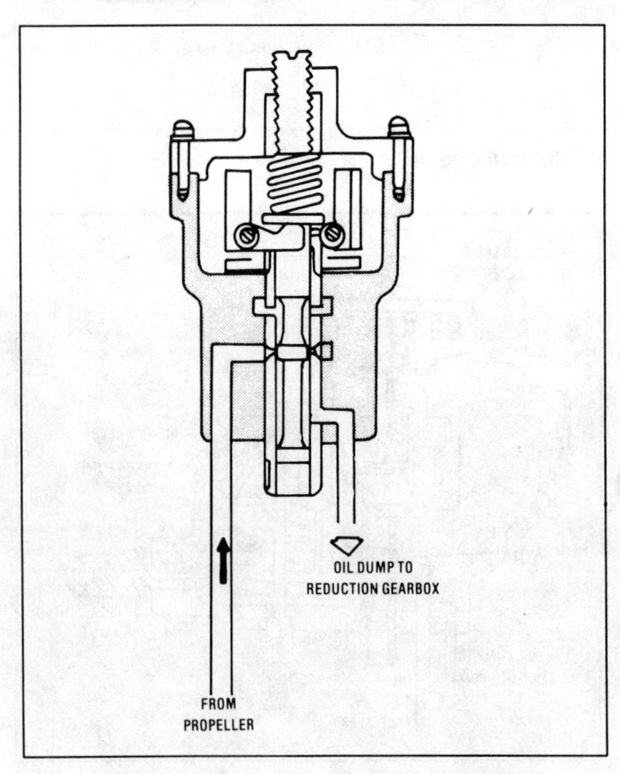

Figure 12-21. Cross section of a PT6 overspeed governor.

The engine fuel control unit is mounted on the rear of the engine and is linked through a cam assembly to the Beta control valve on the propeller governor and to the slip ring on the propeller. This interconnection with the fuel control unit is used during Beta Mode operation.

4. Cockpit Controls

The cockpit controls for the PT6 turboprop installation consist of a power lever controlling engine power output in all modes and propeller blade angle in the Beta Mode, a propeller control lever which adjusts system RPM in the Alpha Mode, and a fuel cut-off lever which turns the fuel on and off at the fuel control unit.

The power lever is linked to the cam assembly on the side of the engine and from there, rearward to the fuel control unit and forward to the propeller governor Beta control valve. The power lever adjusts both engine fuel flow and propeller blade angle in the Beta Mode (reverse to flight idle). In the Alpha Mode, the lever only controls fuel flow to the engine.

The propeller control lever adjusts system RPM in the Alpha Mode through conventional governor operation. Full aft movement of the lever will raise the lift rod in the governor and cause the propeller to feather.

The fuel cut-off lever turns the fuel to the engine on and off at the engine fuel control unit. Some designs have an intermediate position, called *lo-idle*, to limit system power while operating on the ground.

5. System Operation

Beta Mode operation is generally in the range of 50 to 85% RPM. In this range the power lever is used to control both fuel flow and propeller blade angle. When the power lever is moved forward, the cam assembly on the side of the engine causes the fuel flow to the engine to increase. At the same time the linkage to the propeller governor moves the Beta control valve forward out of the governor body, and oil pressure in the propeller is released. As the propeller cylinder moves inboard in response to the loss of oil, the slip ring on the rear of the cylinder moves inboard and, through the carbon block and linkage, returns the Beta control valve to a neutral position. This gives a proportional movement to the propeller.

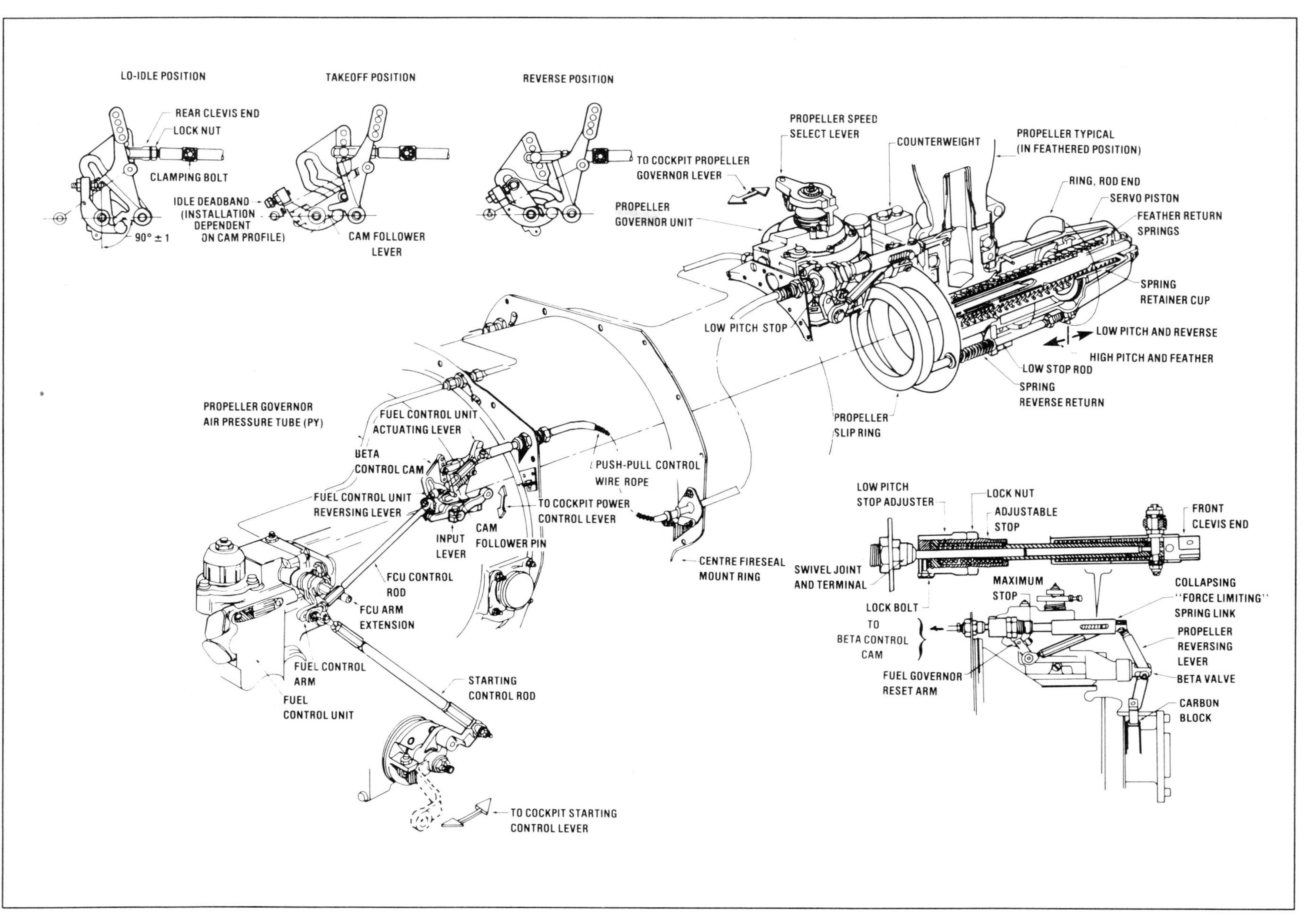

Figure 12-22. Side view of a PT6 engine showing the position of the fuel control, the cam mechanism and the propeller installation.

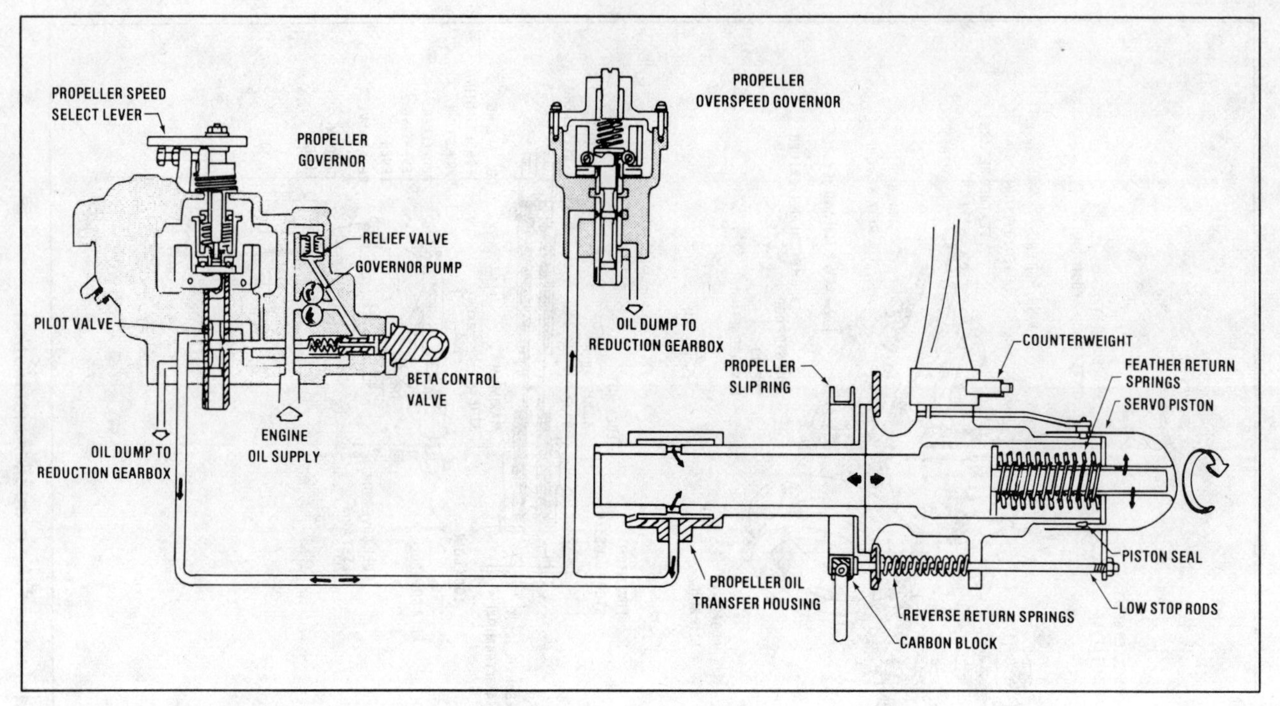

Figure 12-23. PT6 propeller system configuration.

If the power lever is moved rearward, fuel flow is reduced and the Beta control valve is moved in to the governor body directing oil pressure to the propeller to decrease blade angle. As the propeller cylinder moves outboard, the Beta control valve returns to the neutral position by the action of the slip ring, carbon block, and linkage. This again gives a proportional response.

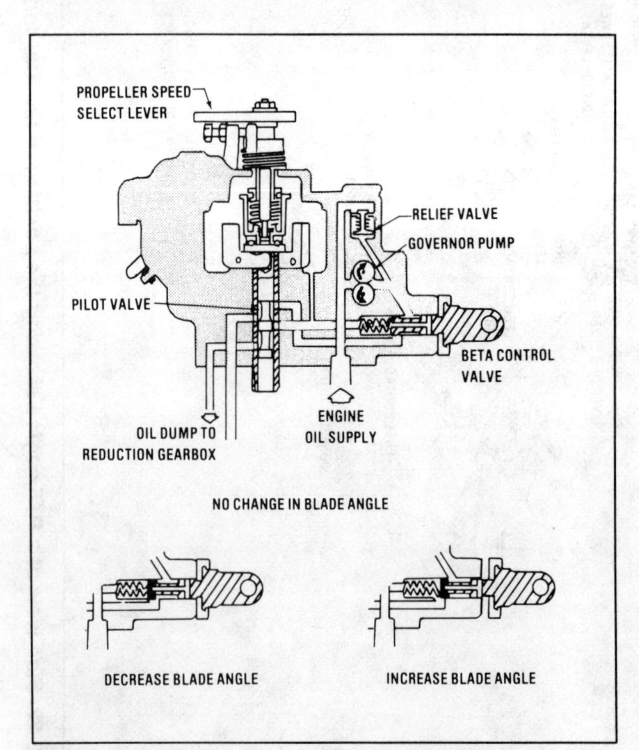

Figure 12-24. Position of the Beta Control valve in the propeller governor to increase, hold and decrease blade angle during Beta Mode.

If the power lever is moved aft of the zero thrust position, fuel flow will increase and the blade angle goes negative to allow a variable reverse thrust. This change in fuel flow is caused by the cam mechanism on the side of the engine.

During the Beta Mode, the propeller governor constant-speed mechanism is underspeed with the pilot valve lowered. The governor oil pump supplies the oil pressure for propeller operation in the Beta Mode.

In the Alpha Mode, the system RPM is high enough for the propeller governor to operate and the system is in a constant-speed mode of operation. As the power lever is moved forward, more fuel flows to the engine to increase horsepower and the propeller governor causes an increase in propeller blade angle to absorb the power increase and maintain the selected system RPM. If the power lever is moved aft, the blade angle will be decreased by the governor to maintain the selected RPM.

To feather the propeller, the propeller control lever is moved full aft, the pilot valve in the governor is raised by a lift rod, and all of the oil pressure in the propeller is released. The springs and counterweights in the propeller will take it to feather.

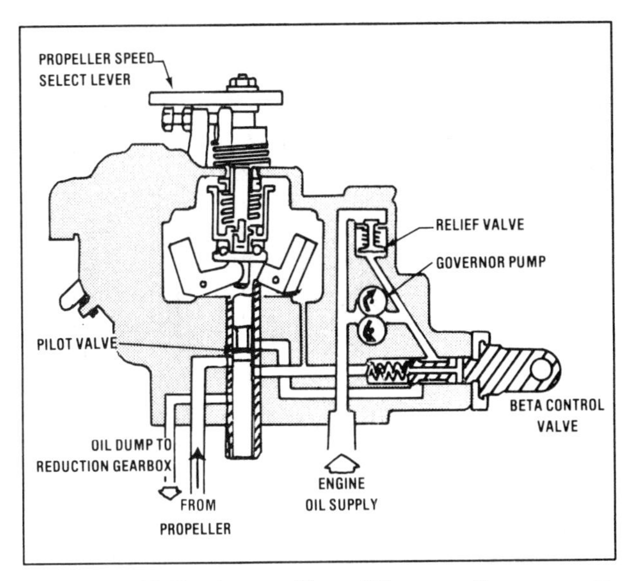

Figure 12-25. Feather position of the propeller governor.

To unfeather the propeller, the engine is started. As it starts to rotate, the power turbine will rotate and the governor or Beta control valve will take the propeller to the selected blade angle or governor RPM setting. When the engine is restarted, the engine will be started before the propeller is rotating at the same proportional speed because of the free turbine characteristic of the engine.

If the propeller RPM should exceed 100%, the propeller overspeed governor will raise its pilot valve and release oil from the propeller to increase blade angle and prevent overspeeding of the propeller. The overspeed governor is automatic and is not controllable in flight.

The power turbine governor prevents excessive overspeeding of the propeller by reducing fuel flow to the engine at approximately 105% RPM. This governor is not controllable in flight and is automatic in operation.

6. Installation And Adjustment

The propeller is installed following the basic procedures for flanged-shaft installations. The slip ring and carbon block arrangement must be installed following the procedures in the aircraft or engine service manual for the specific model involved.

The feather and reverse low blade angles can be adjusted in the same manner as used for the TPE-331 installation.

The governors associated with this system are installed following the procedures in the engine service manual. The governors are then adjusted and rigged with the fuel control unit and the cam mechanism according to the appropriate maintenance manual.

7. Inspection, Maintenance And Repair

The comments for the TPE-331 installation are applicable to the PT6 installation.

8. Troubleshooting

Basic troubleshooting procedures as have been previously discussed, are applicable to the Hartzell reversing system. If the proper propeller response does not occur, check the system for proper rigging before investigating individual units unless the defect is obvious.

In Beta Mode operation, the interconnection should be checked between the power lever, the cam mechanism, the fuel control unit, the Beta control valve on the propeller governor, and the slip ring.

In the Alpha Mode, the propeller governor and linkage to it from the propeller control lever and the cam mechanism should be checked.

If the system RPM is too low, the fault may be with the adjustment of the propeller overspeed governor or the power turbine governor. These components are not involved in the control linkage rigging, but do not forget them. They may also be a reason for overspeeding of the system.

C. Hamilton-Standard Reversing Hydromatic® System

The Hamilton-Standard reversing Hydromatic® propeller system is used on many large transport aircraft such as the Lockheed Electra and the Martin 404. The system uses many of the principles and components of the feathering Hydromatic® models in its construction and operation.

1. Propeller

The reversing Hydromatic® propeller consists of a barrel assembly and a dome assembly that are basically the same as those used in the feathering Hydromatic® propeller, except that the cams and blades are designed for a greater blade angle range. The stop rings on the base of the dome assembly are used to set the reverse and feather blade angles. The low blade angle for constant-speed operation is set by the low-pitch stop-lever assembly which replaces the distributor valve in the center of the dome assembly. The governor performs the function of the distributor valve.

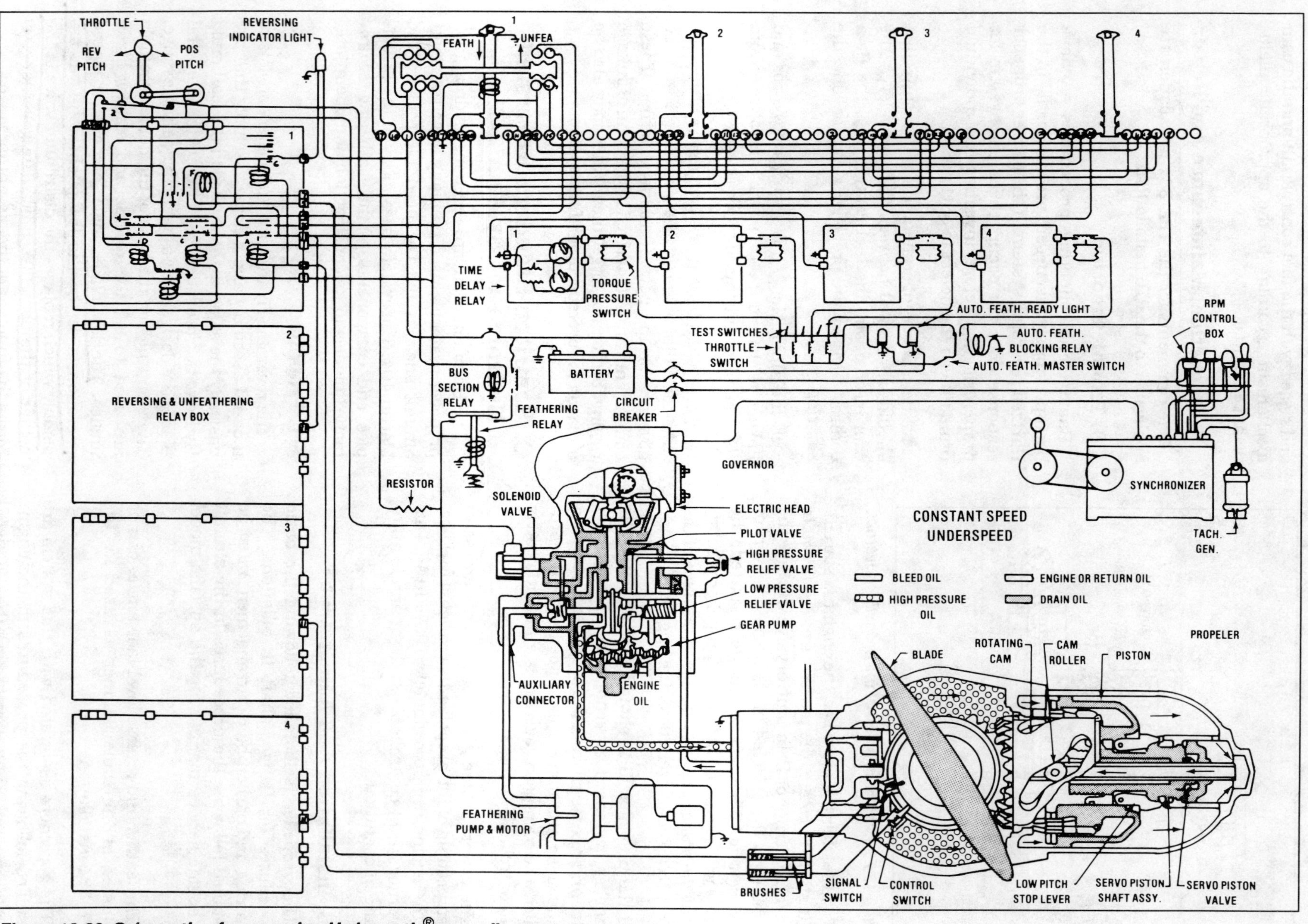

Figure 12-26. Schematic of a reversing Hydromatic® propeller system.

The low-pitch stop-lever assembly in the propeller sets the low blade angle for constant-speed operation and is located in the center of the dome assembly, where it restricts the forward movement of the propeller piston toward a lower blade angle. The stop levers in the assembly are retracted by oil pressure during reversing. They allow the piston to move to the reverse angle stop.

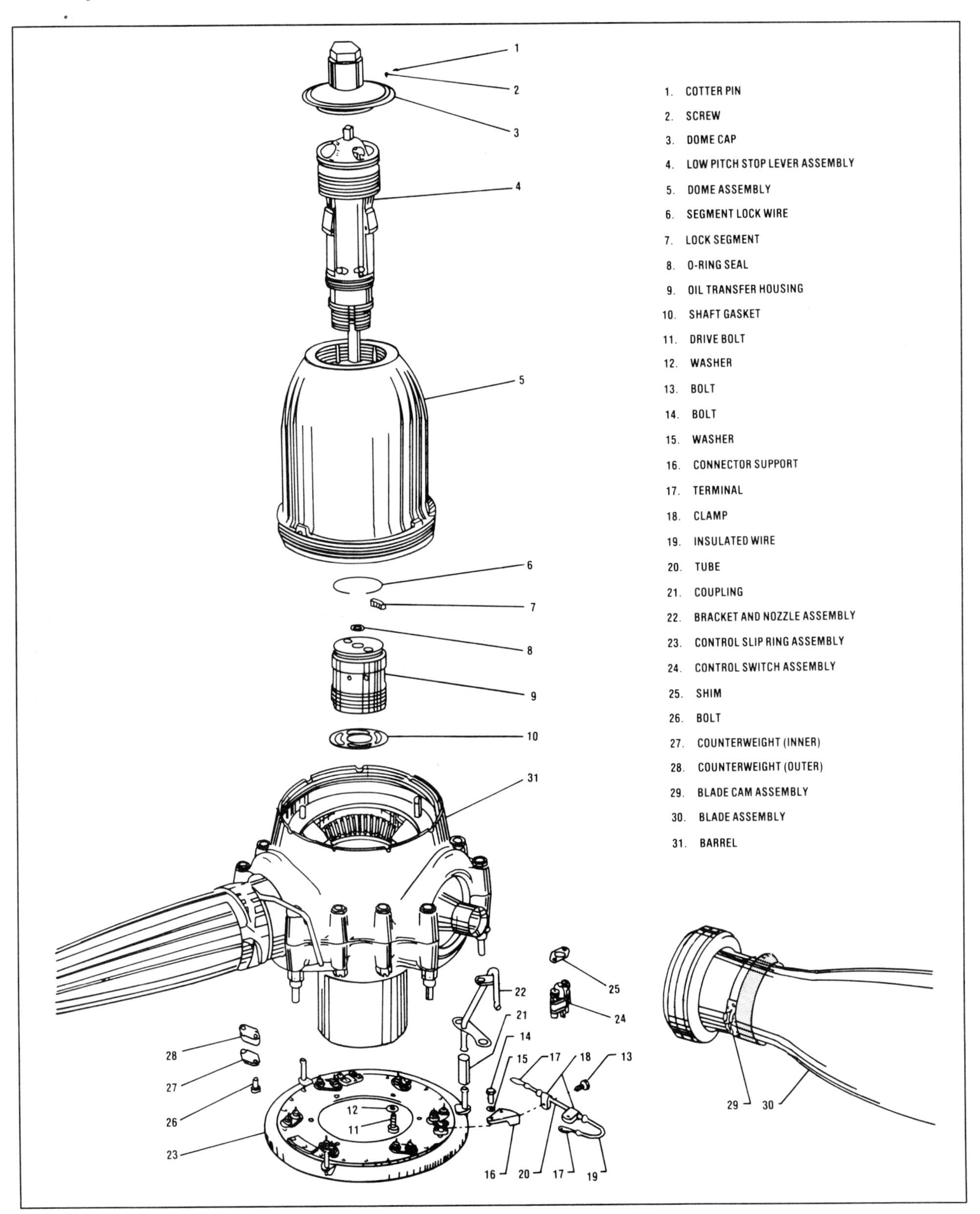

Figure 12-27. Exploded view of a reversing Hydromatic® installation.

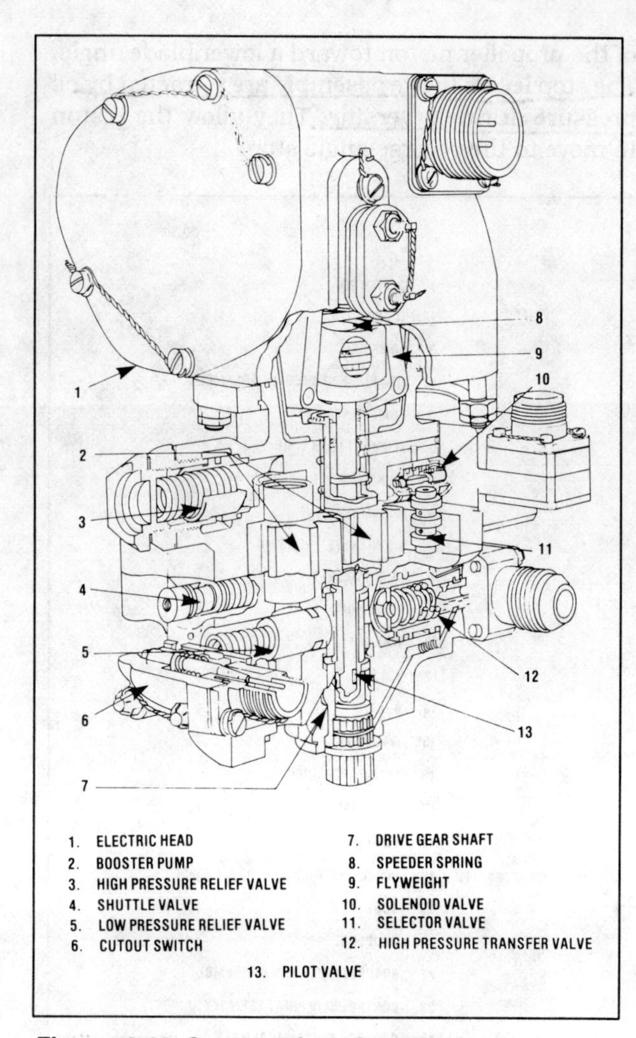

Figure 12-28. Cutaway view of a reversing Hydromatic® governor with an electric head.

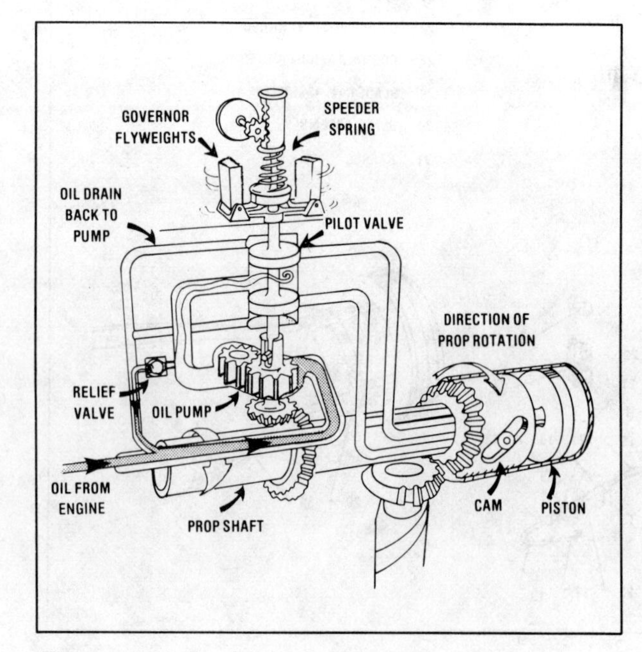

Figure 12-29. Double-acting governor onspeed.

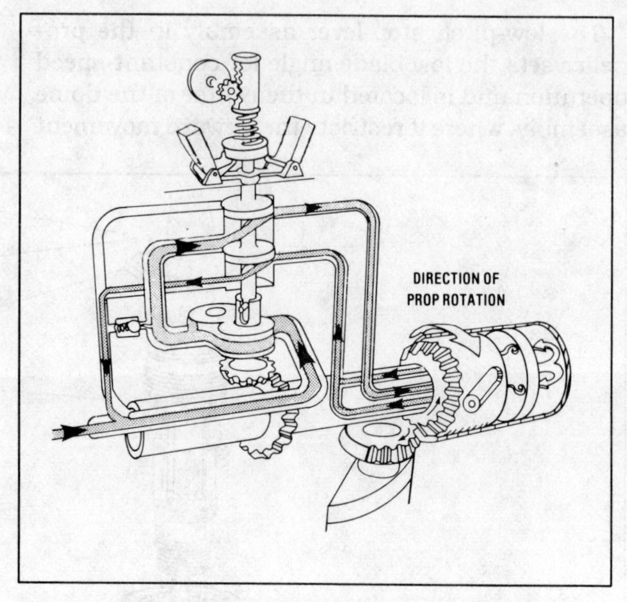

Figure 12-30. Double-acting governor overspeed.

2. Governor

The governor for the reversing Hydromatic® system is similar to the feathering Hydromatic® governor, but contains several additional components and a different style of pilot valve. Most of the reversing governors use electric heads rather than mechanical heads.

The governor is termed a double-acting governor because the pilot valve will direct both engine oil pressure and governor oil pressure to the appropriate side of the propeller piston to change blade angle as shown in Figures 12-30 and 12-31. gov. pump shaft most likely to fail in reverse

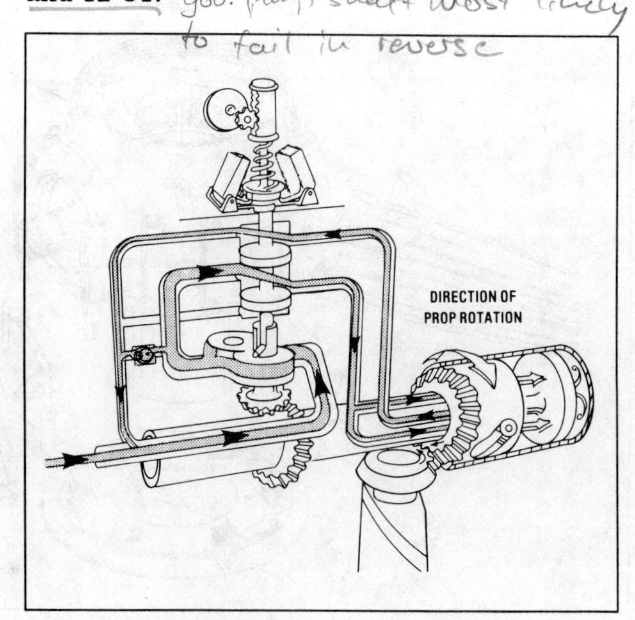

Figure 12-31. Double-acting governor underspeed.

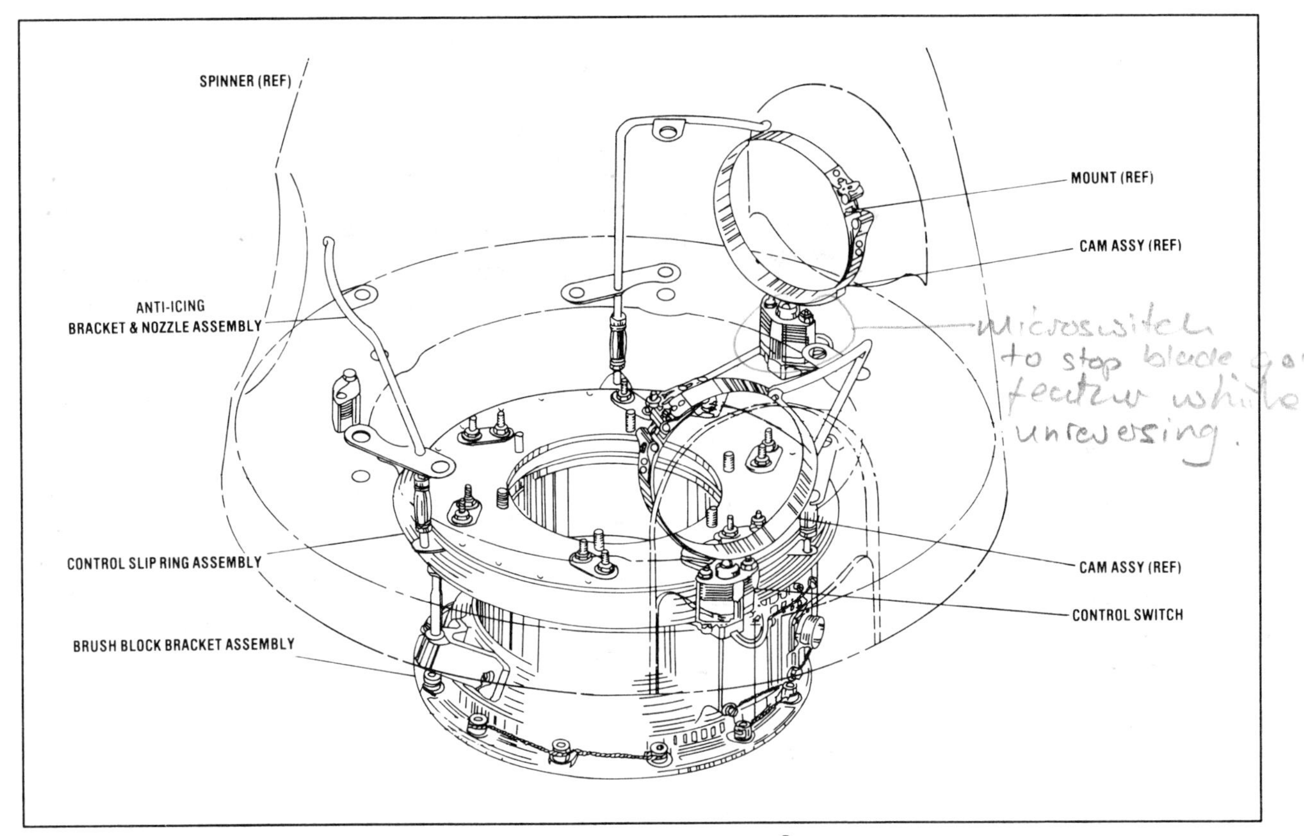

Figure 12-32. Blade switch arrangement on the reversing Hydromatic® propeller.

The high-pressure transfer valve used in the reversing governor, allows oil pressure to enter the governor from the auxiliary pump during feathering, unfeathering, reversing, and unreversing operations and combines this auxiliary oil pressure with the pressure from the governor oil pump during reversing and unfeathering operations to change the blade angles. The governor control mechanism which is used to direct oil flow during these operations, is not cut out of the system as in the feathering Hydromatic® propeller.

A blocking valve is used to block the governor oil pressure relief valve out of the system during reversing and unfeathering operations, thereby allowing the governor oil pump to generate high oil pressures. This valve is positioned by the auxiliary oil pressure and a solenoid operated selector valve. A spring retracts the blocking valve when oil pressure is released by the selector valve when reversing and unfeathering are terminated.

An electric solenoid operated selector valve is used to direct auxiliary oil pressure to the positioning land on the top of the pilot valve during feathering, unfeathering, reversing, and unreversing operations to override the action of the flyweights and speeder spring and hydraulically position the pilot valve as necessary for the desired operation. The selector valve is also used to position the blocking valve during the reversing and unfeathering operations (see Figure 12-37).

The governor contains a pressure cut-out switch which operates the same way that the switch does on a feathering Hydromatic® propeller. The switch is used to terminate the feathering operation and to stop the auxiliary pump when reversing the propeller.

3. System Components

Components, external to the governor and propeller mechanism, are necessary for proper operation.

An auxiliary oil pump is used for all operations other than constant-speed operation and is the same as that used in the feathering Hydromatic® system.

Blade switches and cam rings on the shank of two of the propeller blades are used to terminate the unfeathering and unreversing operations.

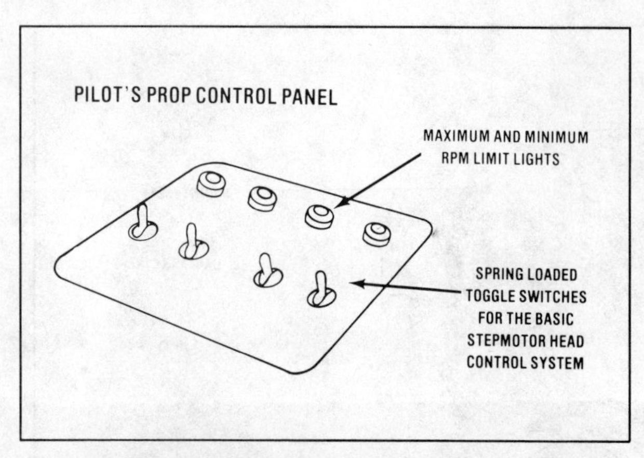

Figure 12-33. Toggle switches are used to control the electric head governors from the cockpit. Lights are used to indicate full movement of the governor speeder rack.

These cam rings are adjustable around the blade shank. The microswitches which are operated by the cam rings, transmit their electrical signal through a slip ring unit mounted on the rear of the propeller barrel. This rotating slip ring transfers the electrical signal to the propeller control system through a ***brush block*** unit mounted on the engine nose case just behind the propeller.

A feathering button is used to feather and unfeather the propeller. Because different electrical circuits are used during feathering and unfeathering, the button is pushed to feather and pulled to unfeather the propeller.

The cockpit propeller control lever is used to set the propeller governor mechanical head for constant-speed operation. If an electric head is used,

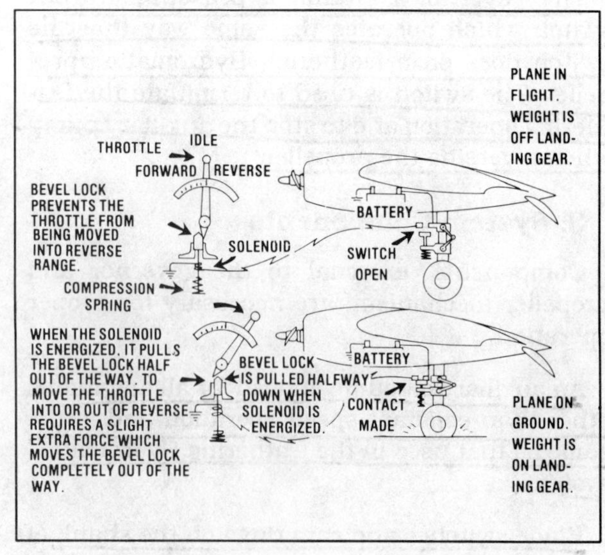

Figure 12-34. Landing gear microswitch operation in the reversing system.

the cockpit control for each propeller is a spring-loaded to center-off toggle switch. The switch is pushed forward to increase the governor RPM setting and moved rearward to decrease RPM. When the speeder rack in the governor electric head has reached full travel, an indicator light will appear next to the toggle switch.

The throttles are used to place the propellers in reverse and to unreverse them. If the throttles are moved aft of the idle position, the propellers will go to full reverse angle. If they are then returned to the idle position, the propellers will unreverse.

A landing gear microswitch, called a squat switch, is used to prevent the propellers from being reversed before the aircraft weight is on the landing gear. This prevents reversing in flight.

Several electrical relays are used to control the propeller feathering and reversing operations. Their use will be illustrated during the discussion of system operation.

4. System Operation

The constant-speed operation of the reversing Hydromatic® propeller system is the same as for other constant-speed systems except that a double-acting governor is used and the position of the fixed force (engine oil pressure) is controlled and shifted by the pilot valve.

To feather the propeller, the feather button is pushed. This completes the circuit to the auxiliary pump and the circuit to the feather button holding coil through the governor pressure cutout switch. The auxiliary pump pressurizes the oil from the oil supply tank and the oil flows to the governor and shifts the high-pressure transfer valve. The ball check valve is seated by this pressure's being greater than that from the governor oil pump. The high oil pressure moves through the selector valve to the positioning land on the top of the pilot valve and forces the pilot valve up into an artificial overspeed condition. This directs the auxiliary oil pressure to the outboard side of the piston and opens the engine oil pressure to the inboard side of the piston.

As the piston moves inboard, the blade angle increases until the stop ring on the base of the dome assembly contacts the stop lug on the stationary cam and the blades stop moving. The blades are then in feather and the oil pressure from the auxiliary pump increases rapidly until the pressure cut-out switch opens and releases

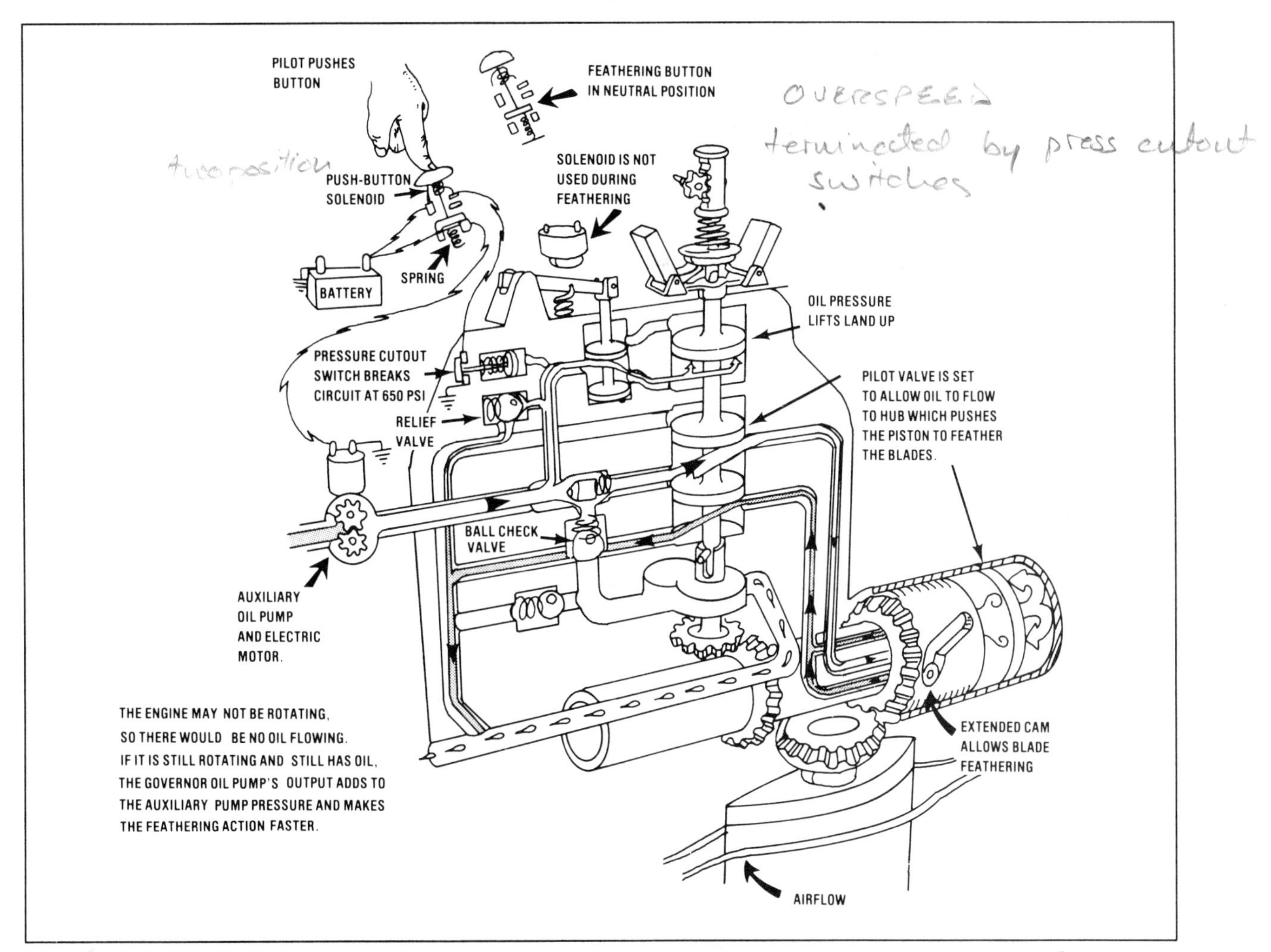

Figure 12-35. A representation of the governor operation during feathering of the reversing Hydromatic® propeller.

the feather button, terminating the feathering operation. The engine is no longer rotating, all pressure drops to zero, and the blades are held in feather by aerodynamic forces (Figure 12-35).

To unfeather the propeller, the feathering button is pulled out. This energizes an electric relay whose holding coil is grounded through a propeller blade switch. The relay completes the circuit to the auxiliary oil pump and energizes the selector valve solenoid. The auxiliary oil pressure enters the governor, moving the high-pressure transfer valve. Oil pressure then flows to the selector valve which is raised. It then flows to the top of the positioning land on the pilot valve, forcing the pilot valve down into an artificial underspeed condition. This directs oil pressure from the auxiliary pump to the inboard side of the piston and moves the piston outboard, causing the blade angle to decrease and the propeller will start to windmill. The engine is restarted once the feather button is released (Figure 12-36).

If the feather button is not released, the blade microswitch and blade cam ring will break the electric circuit to the holding coil on the relay between the feather button and the auxiliary pump. This microswitch terminates the unfeathering operation at five to seven degrees before the piston reaches the low blade angle stop. This occurs if the feather button is not released by the crewmember when the engine starts to windmill.

To reverse the propellers, the aircraft must be on the ground with the aircraft weight on the landing gear, allowing the squat switch to close. This removes a lock on the throttles and allows them to be moved rearward into the reverse range. When the throttle is moved rearward, an electric circuit is completed to the solenoid selector valve and the auxiliary pump. The auxiliary pump circuit is completed through contacts on the pressure cut-out switch. The auxiliary oil pressure positions the governor components in

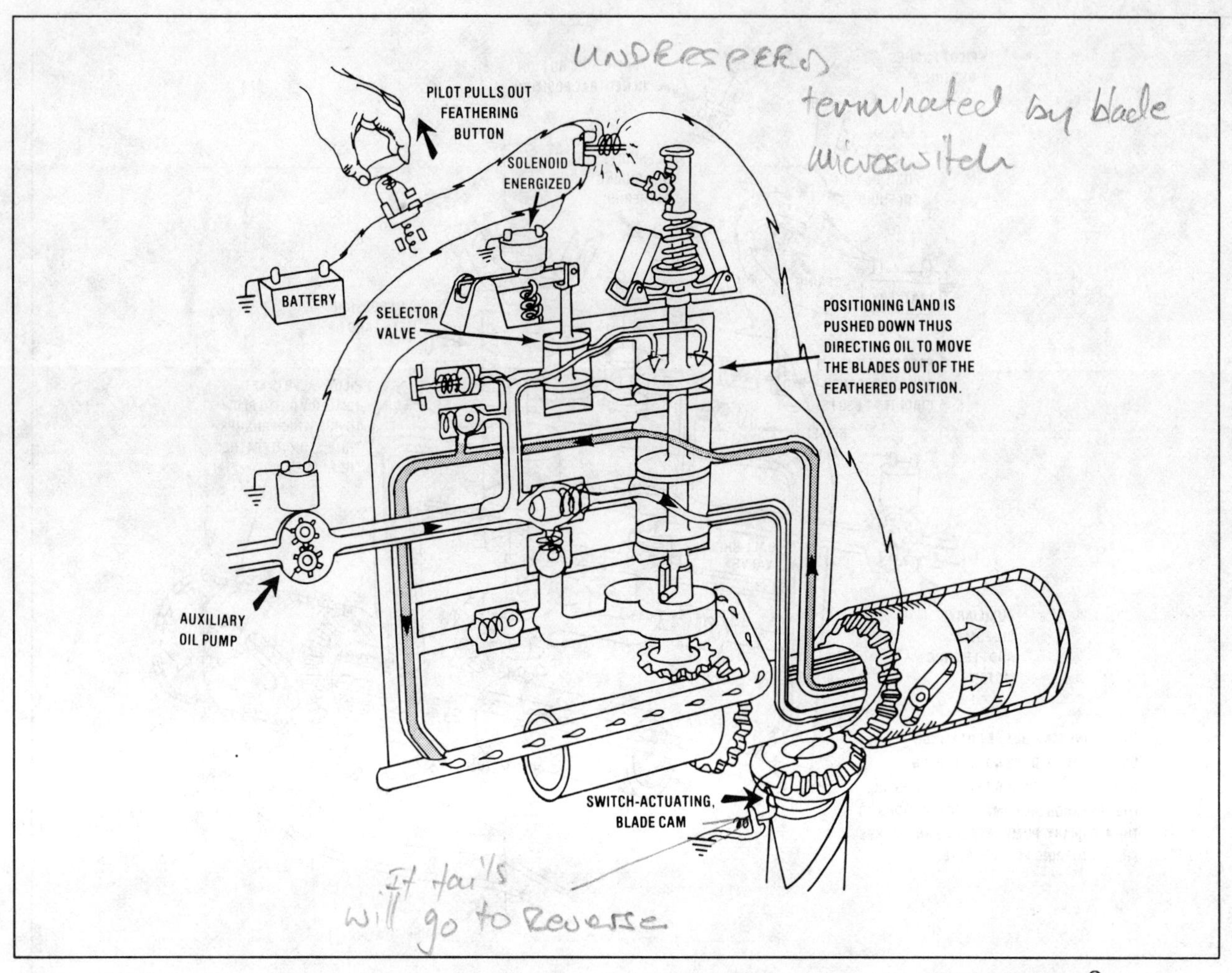

Figure 12-36. A representation of the governor operation during unfeathering of the reversing Hydromatic® propeller.

the same position as for unfeathering the propeller. (Reversing and unfeathering operations both decrease the blade angle (Figure 12-37)).

Note that the blocking valve is inserted to prevent the governor oil pressure relief valve from relieving and the governor oil pressure pump will work with the auxiliary pump to help reverse the propeller.

The oil pressure moves the propeller piston forward to a lower blade angle. Oil pressure is also directed to the servo piston in the low pitch stop-lever assembly, moving the wedge forward and allowing the stop levers to be retracted (Figure 12-38). The piston then moves forward over the stop lever assembly to the reverse angle. When the stop ring contacts the stop lug on the stationary cam, the piston stops moving and the blades are in the reverse angle. The auxiliary pump is shut off by the action of the pressure cutout switch and the propeller is held in reverse by the high pressure being generated by the governor oil pump. (The blocking valve and selector valve are still in position and the high pressure is maintained by the governor high pressure relief valve).

As the throttle position is varied in the reverse range, the engine RPM changes to vary the amount of reverse thrust generated. (The farther aft the throttle is moved the higher the reverse RPM.) In reverse the blade angle does not vary, but stays at the reverse angle limit.

To unreverse the propeller, move the throttle forward to the idle position. This restarts the auxiliary pump through the pressure cut-out switch and a blade microswitch. Auxiliary oil pressure enters the governor and raises the pilot valve through the selector valve positioning land action. The governor is in the same configuration as for feathering the propeller. (Both operations increase the blade angle.) Oil pressure moved the piston inboard and the stop levers are re-inserted as the low blade angle stops as the piston moves inboard of the levers. When the

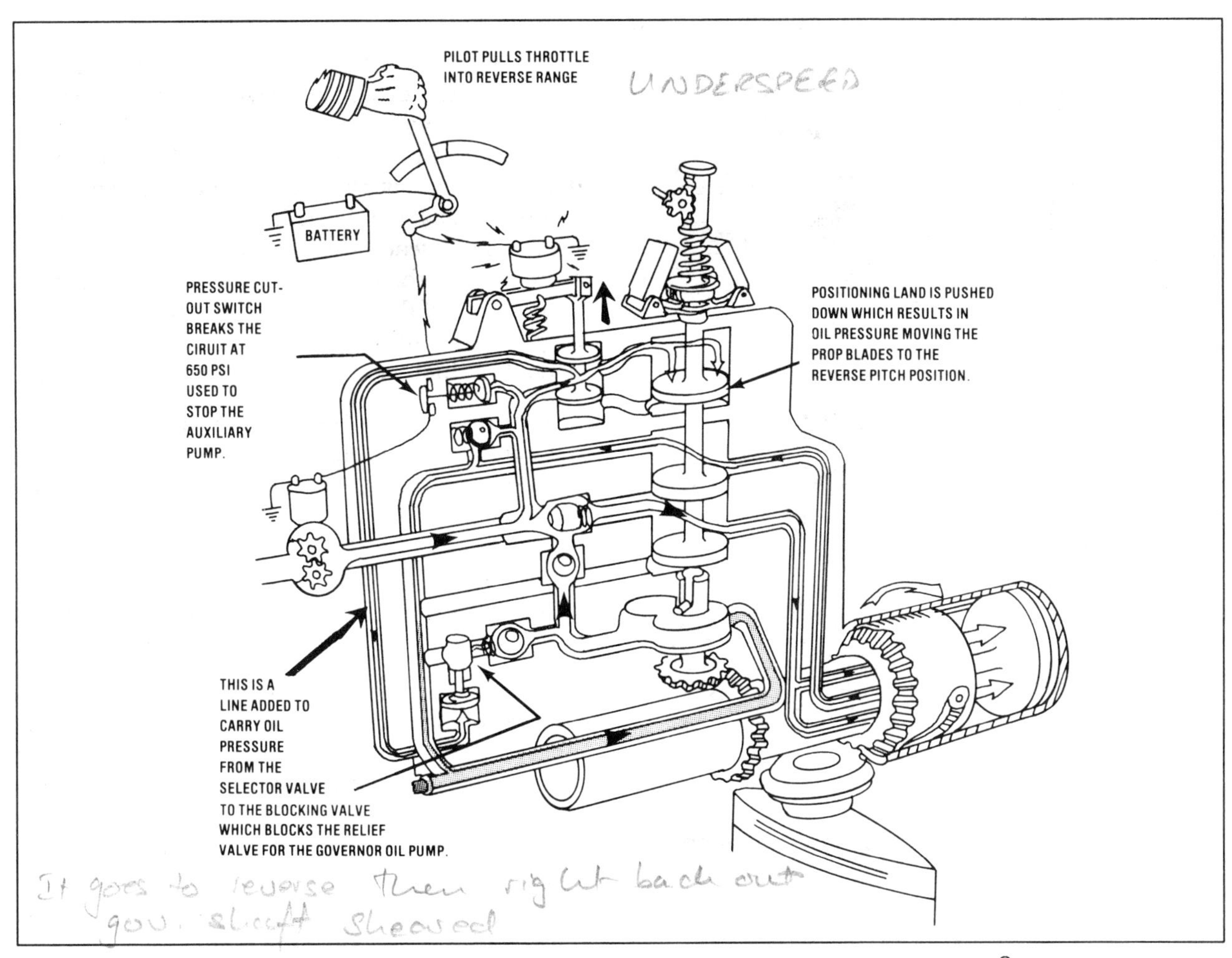

Figure 12-37. A representation of the governor operation during reversing of the Hydromatic® propeller.

blades rotate to an angle a few degrees greater than the low blade angle setting, the microswitch and cam on the blade terminate the unreversing operation and the system returns to constant-speed operation (see Figures 12-39 and 12-40).

5. Installation And Adjustments

The propeller barrel assembly and dome assembly are installed following the same procedure as for the feathering Hydromatic® propeller. At this time, the reverse and feather blade angles should be checked.

The low pitch stop-lever assembly is then installed through the front of the dome. The blades should be at some high blade angle during this installation to prevent interference with the stop lever assembly. The assembly is screwed into the dome to a predetermined depth, for the particular aircraft, to set the low blade assembly. Check the low blade angle by rotating the blades until the piston contacts the stop levers. Measure this angle and turn the stop lever assembly in to increase the low blade angle, or out to decrease the low blade angle. The assembly is safetied and the dome cap is installed and safetied.

The governor is installed and adjusted following the procedures covered for other Hamilton-Standard governors. Connect and safety all of the cannon plugs required for the installation.

Propeller controls and equipment are installed following the procedure in the maintenance manual for the particular aircraft model.

6. Inspection, Maintenance And Repair

Inspection, maintenance, and repair of the reversing Hydromatic® propeller is basically the same as for the feathering Hydromatic® system.

7. Troubleshooting

For constant-speed and feathering troubleshooting, refer to the section on feathering Hydromatic® propellers.

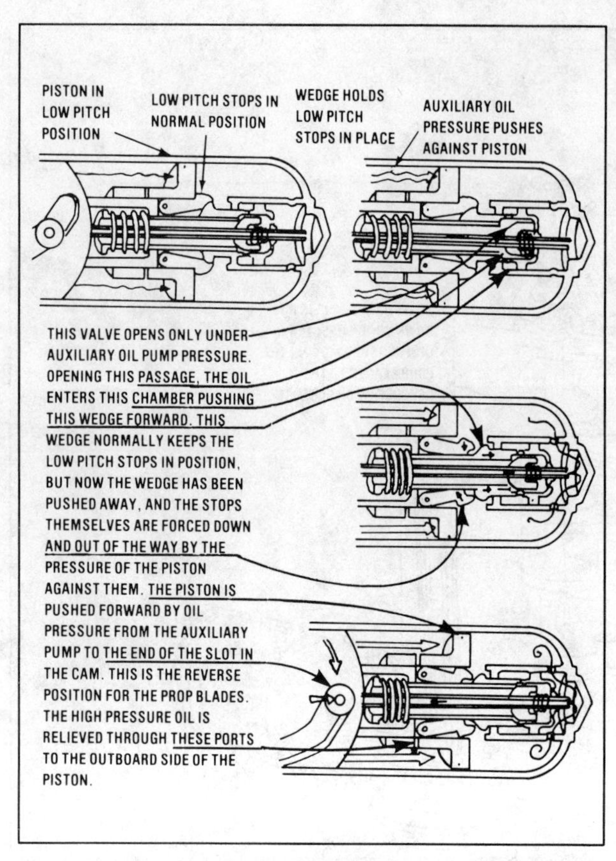

Figure 12-38. Operation of the stop-lever assembly during reversing of the Hydromatic® propeller.

If the propeller will not unfeather, one of the following may be the problem: the selector valve may not operate; the blade switch may be open; the blade cam may be positioned incorrectly; the unfeathering contacts on the feather button may be defective; the slip ring or brush block behind the propeller may not be transmitting the electrical signal; the unfeathering relay may be defective. The correction for these defects involves replacement or adjustment of the defective component. The defect may also be the wiring associated with the components rather than the components themselves.

If the propeller will not reverse, one of the following may be the problem: the squat switch may be open or improperly adjusted; the solenoid selector valve may be defective; the low pitch stop-lever assembly may not be retracting the wedge from behind the stop-levers; the pressure cut-out switch may be open; the throttle contact may be open. The defect should be corrected as appropriate.

If the propeller will not unreverse, the defect may be: a defective relay; defective blade microswitch; incorrectly positioned blade cam; slip ring or brush block difficulty; incorrect setting of the stop-lever assembly where the blade switch opens before the piston moves inboard past the stop levers. Correct the defect as appropriate.

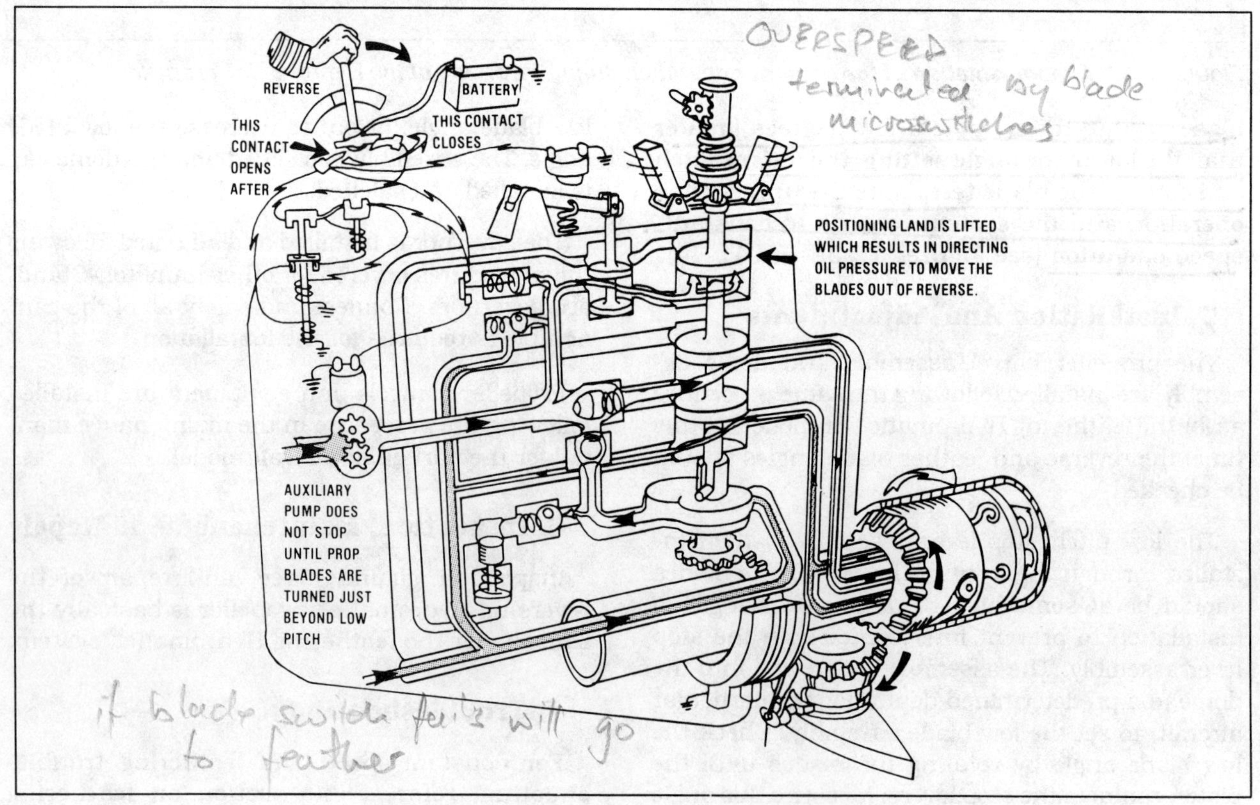

Figure 12-39. A representation of the governor operation during unreversing of the Hydromatic® propeller.

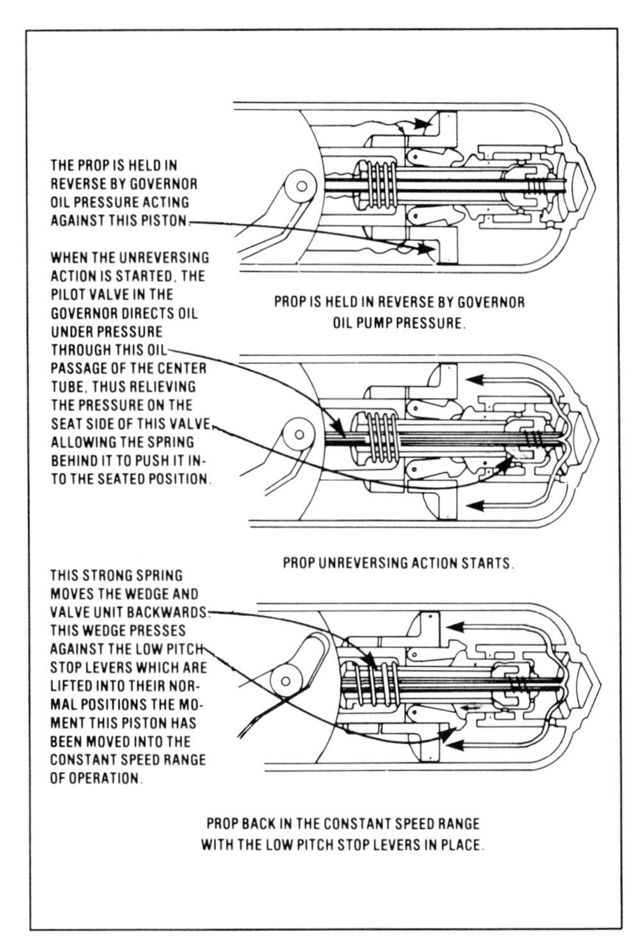

Figure 12-40. Operation of the stop-lever assembly during unreversing of the Hydromatic® propeller.

QUESTIONS:

1. *What are some advantages of using reversing propellers?*
2. *What is the main disadvantage of using reversing propellers on reciprocating engine powered aircraft?*
3. *Which cockpit control is used to control the reversing operation?*
4. *Normally, what condition must exist before the propellers can be reversed?*
5. *What force(s) is/are used to increase propeller blade angle with a Hartzell Steel Hub reversing propeller?*
6. *For the TPE-331 installation, which engine accessories are used to control system RPM?*
7. *For the TPE-331 installation, which engine accessories are used to control propeller blade angle?*
8. *What does Beta Mode of operation indicate?*
9. *What does the feathering valve do when it is moved in the TPE-331 installation?*
10. *What does the condition lever adjust for the TPE-331 installation in the Beta Mode? In the Alpha Mode?*
11. *Does the TPE-331 propeller or the PT6 propeller normally shutdown in the feather position?*
12. *In which component of the PT6 reversing system is the Beta control valve located?*
13. *What governor component in the PT6 system is used to feather the propeller in flight?*
14. *What does the power lever control in the PT6 system during Beta Mode operation?*
15. *What does the power turbine governor control in the PT6 installation?*
16. *What is the purpose of the low pitch stop-lever assembly in the reversing Hydromatic® system?*
17. *What angles are set by the stop rings in the reversing Hydromatic® propeller?*
18. *Which system pump is used to keep the reversing Hydromatic® propeller in reverse?*
19. *What is the purpose of the squat switch in the reversing Hydromatic® propeller?*
20. *As far as the governor is concerned in the reversing Hydromatic® system, reversing is the same as?*

Chapter XIII
Propeller Auxiliary Systems

Propeller auxiliary systems include systems which increase the efficiency of propeller operation and provide automatic operation of governor and feathering mechanisms to reduce the fatigue and workload of the flight crew. One or more of the systems discussed in this chapter may be found on almost any aircraft from light single-engine airplanes to large transports.

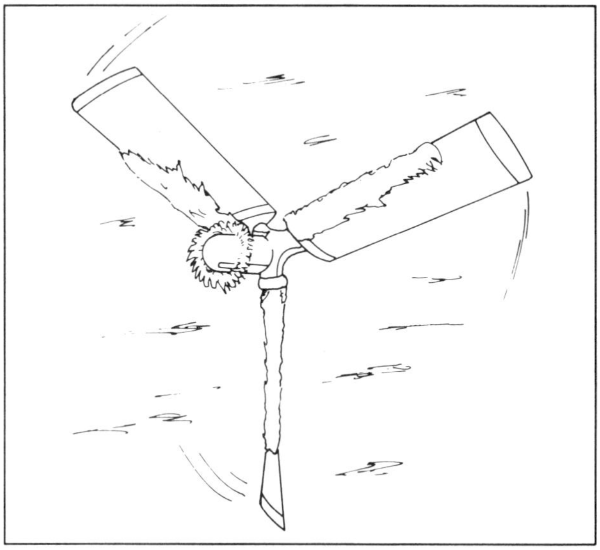

Figure 13-1. Ice accumulation on a propeller.

A. Ice Elimination Systems

Propeller ice elimination systems are used to prevent or remove ice formation on propeller blades during flight. If ice is allowed to remain on the blades, the efficiency of the airfoil is reduced, the propeller becomes heavier, and develops an out-of-balance condition. These conditions can generate vibrations and cause damage to the engine and the airframe.

Two types of ice elimination are used — ***anti-icing*** and ***de-icing***.

1. Anti-Icing

Anti-icing refers to any system which prevents the formation of ice on the propeller. The most commonly used type of anti-icing system employs a fluid which mixes with the moisture on the

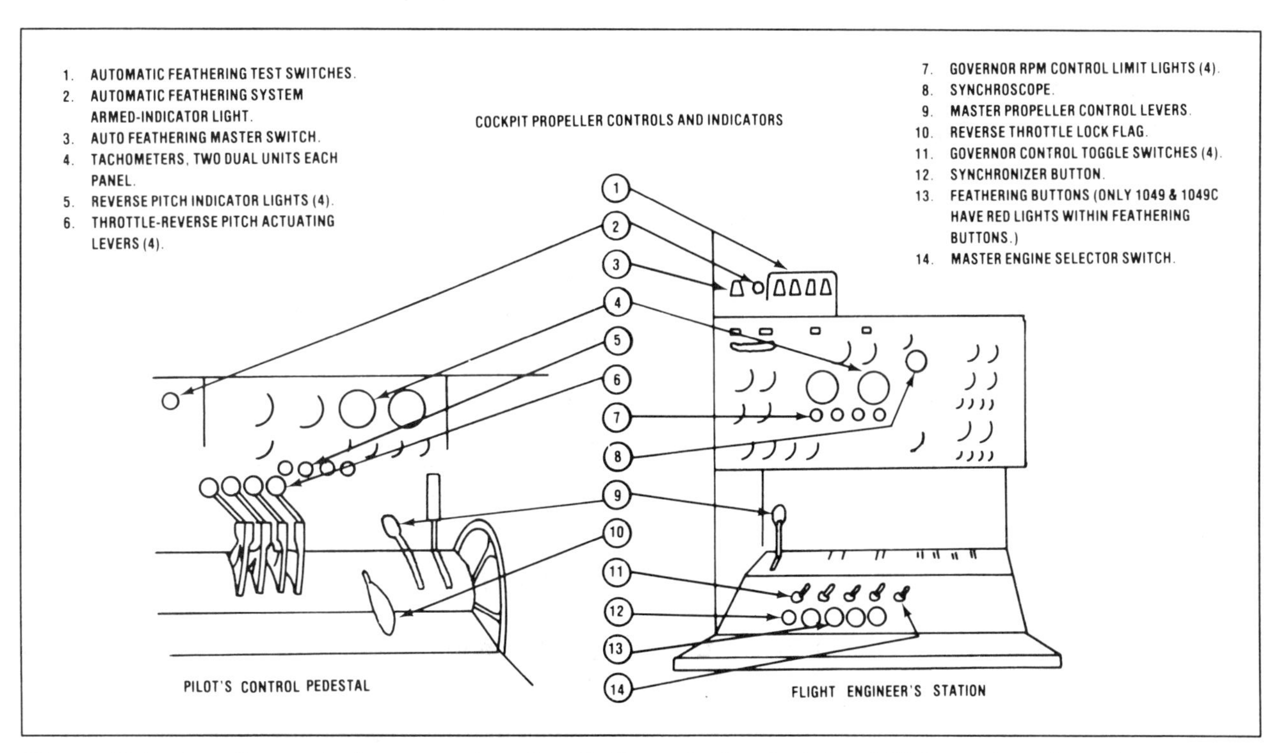

Figure 13-2. Propeller controls mounted on the pilot's control pedestal and flight engineer's station on a Lockheed Constellation.

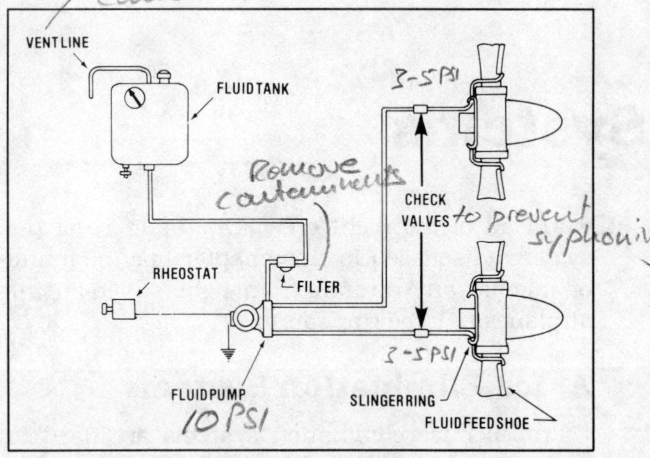

Figure 13-3. A typical propeller fluid anti-icing system.

propeller blades and allows the mixture to flow off of the blades before the moisture can create an ice build-up on the blades. This system is ineffective once the ice has formed, so the system must be in operation whenever the aircraft is operating in suspected icing conditions.

Anti-icing systems are used on aircraft as small as a Cessna 310 and as large as a Douglas DC-6.

a. System Components

The fluid used in the anti-icing system must readily combine with water and have a very low freezing point so that the mixture of fluid and water will not freeze during flight. The most commonly used fluid is isopropyl alcohol because of its low cost and availability. A primary disadvantage is the flammability of the fluid. Another fluid of phosphate compounds is used in some systems, but it is not as widely used due to its high cost, even though it is less flammable than alcohol.

A fluid tank used with the system is usually located in the fuselage and may or may not be accessible in flight, depending on aircraft design. The tank is vented to the atmosphere and contains a quantity indicator. The indicator may be a direct reading or a remote indicating type, as necessary so that the quantity is indicated in the cockpit. The tank is positioned so that it will gravity feed to the fluid pump(s). The size of the tank depends on the aircraft. It may have a capacity of a few quarts to several gallons.

A fluid filter is placed in the line between the tank and the fluid pump to prevent contaminants from entering the system from the tank.

A fluid pump is used to move the fluid from the tank to the propeller feed lines. The pressure

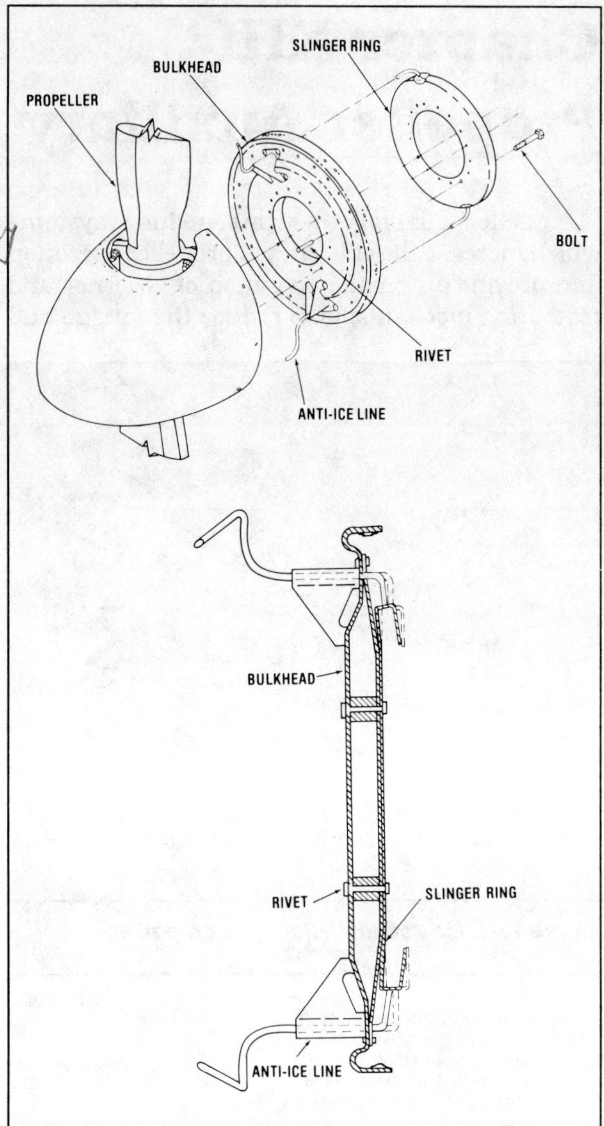

Figure 13-4. Slinger ring installation on a light twin.

developed by the pump is no more than about ten psi as there is very little resistance to fluid flow other than a check valve which opens at three to five psi. The pump speed is controlled from the cockpit by a rheostat and can be varied from less than a quart per hour to more than a gallon per hour of fluid flow. Usually one pump will supply no more than two engines on an aircraft.

A check valve located between the fluid pump and the ***slinger ring*** feed tube is used to prevent siphoning of fluid in flight when the system is not operating and to reduce evaporation of fluid from the system.

A slinger ring feed tube mounted on the engine nose case directs the fluid flow into the slinger ring which is rotating with the propeller.

The slinger ring mounted on the rear of the propeller hub holds fluid in its curved channel by centrifugal force. The fluid flows out to the blades through the blade feed tubes, which are outlets welded onto the blade slinger ring.

Rubber feed ***shoes*** (anti-icing ***boots***) which are attached to the leading edge of the propeller blades by an adhesive are optional items and are not used on all systems. The shoes direct the fluid flowing along the leading edge of the blades as it comes out of the feed tubes and provide an even distribution of the fluid. The shoes often do not extend beyond one-third of the blade length.

b. System Operation

When the system rheostat is turned on, the fluid pump operates at the rate set on the rheostat by the pilot. Fluid is drawn from the tank, through the filter, and is forced out to the slinger feed tube. Fluid flows from the stationary feed tube to the rotating slinger ring on the rear of the propeller where it flows through blade feed tubes to the leading edge of the blades at the shank. The fluid flows out of the tube onto the blade surface or boot and moves along the length of the blade leading edge by centrifugal force. The fluid combines with the moisture and the mixture flows off the blades as a liquid.

c. Inspection, Maintenance And Repair

The fluid level in the tank should be checked and serviced before each flight during which the system may be used. Due to the nature of alcohol, the fluid level may be decreased by evaporation over the period of a few days if the conditions are right (high temperature, high altitude, etc.). The system should be serviced with the recommended fluid.

The filter should be cleaned at every 100-hour inspection and annual inspection and as necessary to assure proper operation.

All lines should be checked for condition, security, and obstructions, especially for lines clogged by insects.

The fluid feed shoes should be inspected for deterioration of the shoe material, damage to the shoes from sand, stones, etc., and for separation of the shoe from the blade. Inspect the area around the shoes for any indication of corrosion which may extend under the shoe. If corrosion is found, the shoe should be removed following the aircraft manufacturer's instruction, the area of corrosion treated by the mechanic or an overhaul facility, as appropriate to the extent of corrosion, and a new shoe installed following the manufacturer's instructions.

The system can be checked for proper operation on the ground by first connecting a hose to the feeder tube on the engine nose case and placing it in a calibrated container. Determine the rate of flow that should occur and determine the amount of fluid that should be pumped through the system during some small time interval, such as five minutes. As an example: if the system should supply three gallons of fluid in one hour, in five minutes the system should put out one quart of fluid. If the amount of fluid pumped is substantially less than that desired, the system should be checked for defective components (see the troubleshooting section). The amount of fluid pumped may be slightly less than the amount calculated as the pump is operating from battery voltage which may be 10% less than generator voltage.

d. Troubleshooting

Troubleshooting of the fluid anti-icing system is fairly straightforward. If the system does not produce any fluid at the blades, any component of the system may be the defective unit. The best way to check the system is to determine if fluid is being delivered by the pump. Loosen a fitting at the pump outlet and note if fluid comes out around the loose fitting when the system is turned on. If it does, the defective component is downstream of the pump and if not, the defective component is upstream of the fitting used to check for fluid flow. Investigate further to locate the defective component.

If the fluid is not being delivered to one propeller, one of the following may be the problem: a line may be blocked; the check valve may be stuck closed.

If one blade does not anti-ice, the problem is a blocked line between the slinger ring and the blade feed tube outlet.

If the system fluid supply decreases in flight when the system is not being used, check the system for leaks and check to be sure that the check valves are closed. Correct the leaks as appropriate and repair or replace the check valve.

If the system works properly when it is first turned on, but the flow rate generally decreases and then stops and the fluid supply is sufficient and the pump is operating, the tank vent may be blocked. It will allow a vacuum to build up in the tank as fluid is being used until the fluid will no longer gravity flow to the pump, and flow stops.

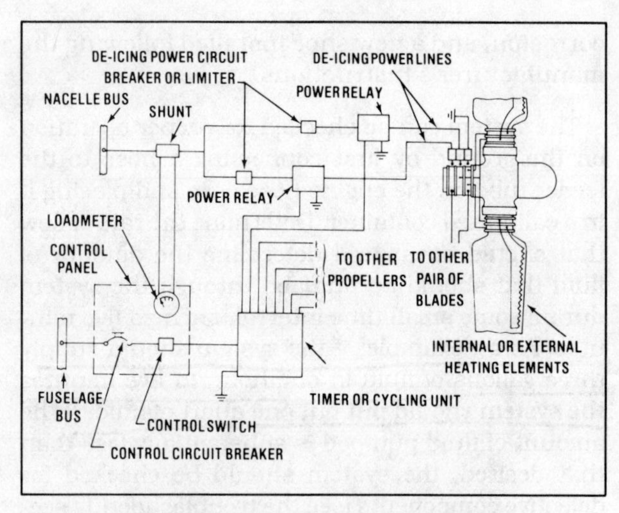

Figure 13-5. Typical electrical de-icing system.

2. De-Icing

De-icing refers to a system which allows ice to form, and then removes the ice from the propeller blades. De-icing systems use electrical heating elements on the blades to melt the layer of ice next to the blade and centrifugal force throws the ice off of the blade. The system heats the blades for a short period of time and then the current is turned off and ice again forms on the blades. Once the ice forms, the system again heats and melts the accumulated ice.

Electrical de-icing is the preferred method of ice control for propellers due to the infinite supply of electrical power during a flight.

This system may be used on aircraft ranging in size from Cessna 210s to Lockheed Electras.

a. System Components

The pilot controls the operation of the de-icing system from the cockpit through one or more toggle switches. An on-off switch is included in all systems to supply power to the system for operation. The system will de-ice as long as the switch is on.

Some systems incorporate a selector switch which can select one of two cycling speeds to adjust for heavy or light icing conditions. Another control switch which may be used is a Full De-ice Mode toggle switch which is spring-loaded off and must be held on to de-ice all of the propellers at the same time. This switch can only be used for short periods of time and is used if ice builds up on the propeller before the system is turned on.

A loadmeter located in the cabin near the system control switches is used to indicate the amount of

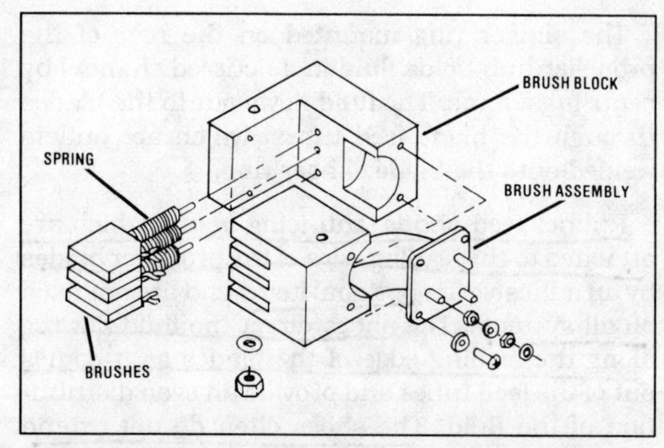

Figure 13-6. Brush block assembly used in the de-icing system of a light twin.

current drawn by the de-icing system. One loadmeter may be used for all engines on a multi-engine aircraft or a separate one may be used for each engine. The loadmeter may be calibrated in amps or in percent of rated current.

The system timer is turned on and off by the cockpit controls and is used to sequence the operation of the de-icing system. A DC motor in the timer runs a sequencing mechanism which turns each propeller de-icing circuit on and off in proper sequence and keeps the proper heating interval for each propeller. At any one instant, only one propeller will be de-iced. The timer is mounted in the aircraft fuselage and controls the de-icing operation through power relays located in the engine nacelles. Thus the length of high current carrying wires is kept to a minimum.

A brush block mounted on the engine nose case just behind the propeller contains brushes which are used to transfer electrical power from the power relays to the propeller slip ring. The slip ring mounted on the rear of the propeller contains at least two contact bars which align with the brushes in the brush block and accept the electrical power from the brushes transferring it to the blade boots. The slip ring is trued for contact ring roundness and is flat to provide even and continuous contact during operation.

Rubber de-icing boots, containing the heating elements, are attached to the blade leading edge with adhesives. Some boots contain two heating elements which reduce the amount of current required during the de-icing operation and allow a greater concentration of heat on the blades. The two elements operate on the inboard and the outboard sections of the boots and are heated in sequence by the action of the timer.

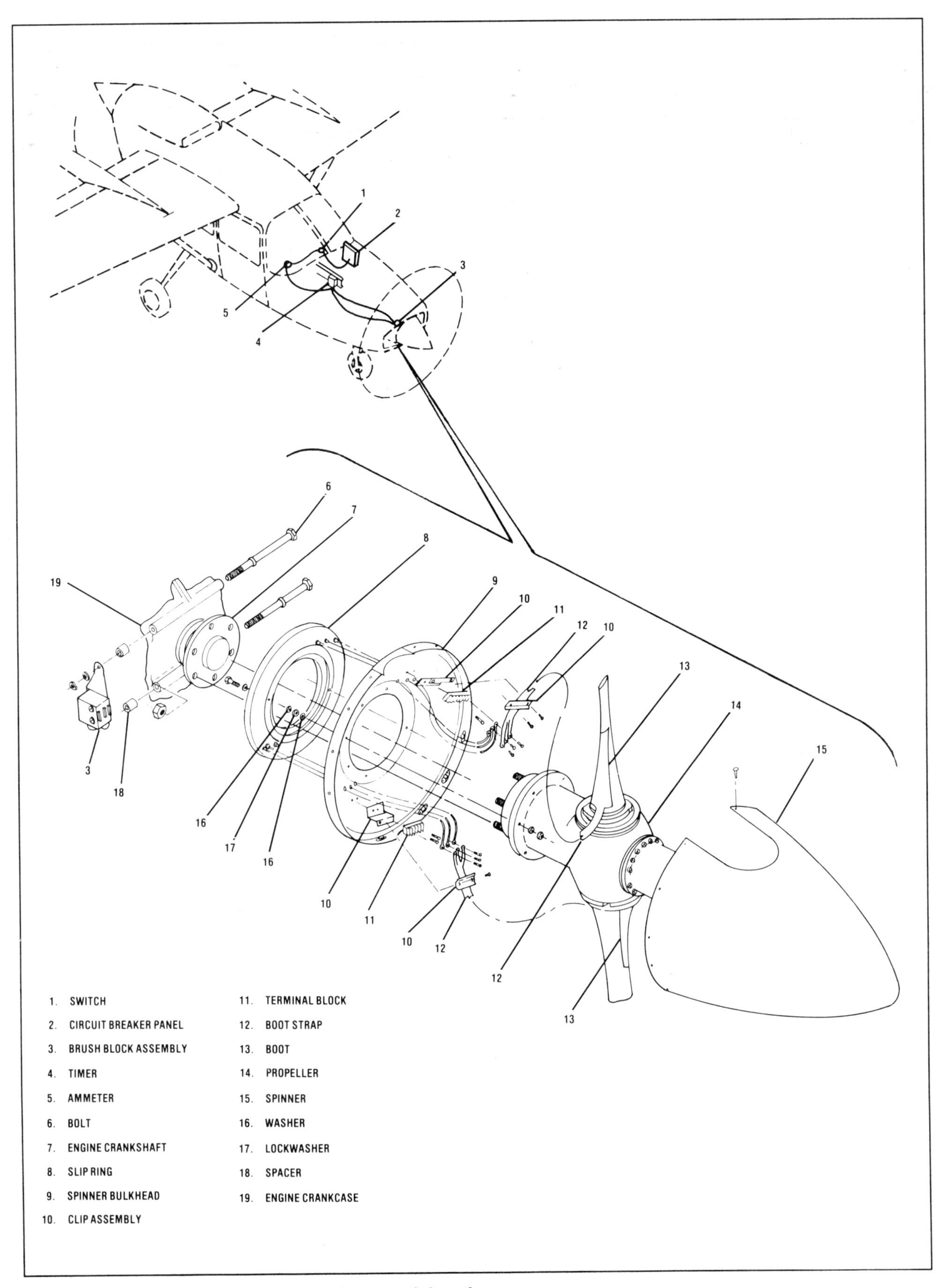

Figure 13-7. De-icing propeller installation on a light twin.

b. System Operation

When the pilot turns on the cockpit switch, the timer starts to run, sequencing the de-icing operation. As the timer sequences, power is delivered to each power relay in turn. The high current released by the power relay is directed to the blade boots through the brush block and slip ring assembly. Each propeller is de-iced in turn by the operation of the timer.

To prevent more than one propeller from de-icing at any one time during normal operation, a null period of about one second is set in the timer. During the null period no blades are being de-iced and the loadmeter indicates zero current flow.

As the timer operates, an indication is given in the cockpit by the action of the loadmeter. If one loadmeter is used, the meter will indicate a current flow for the heating time (approximately 30 seconds) and will then return to zero for about. one second (for the null period) before returning to the current flow reading. If a loadmeter is used for each engine, only one meter will indicate current flow at any one instant and during the null period all meters will indicate zero.

The time for the heating sequence may vary for each model of aircraft, but is about 30 seconds for each propeller.

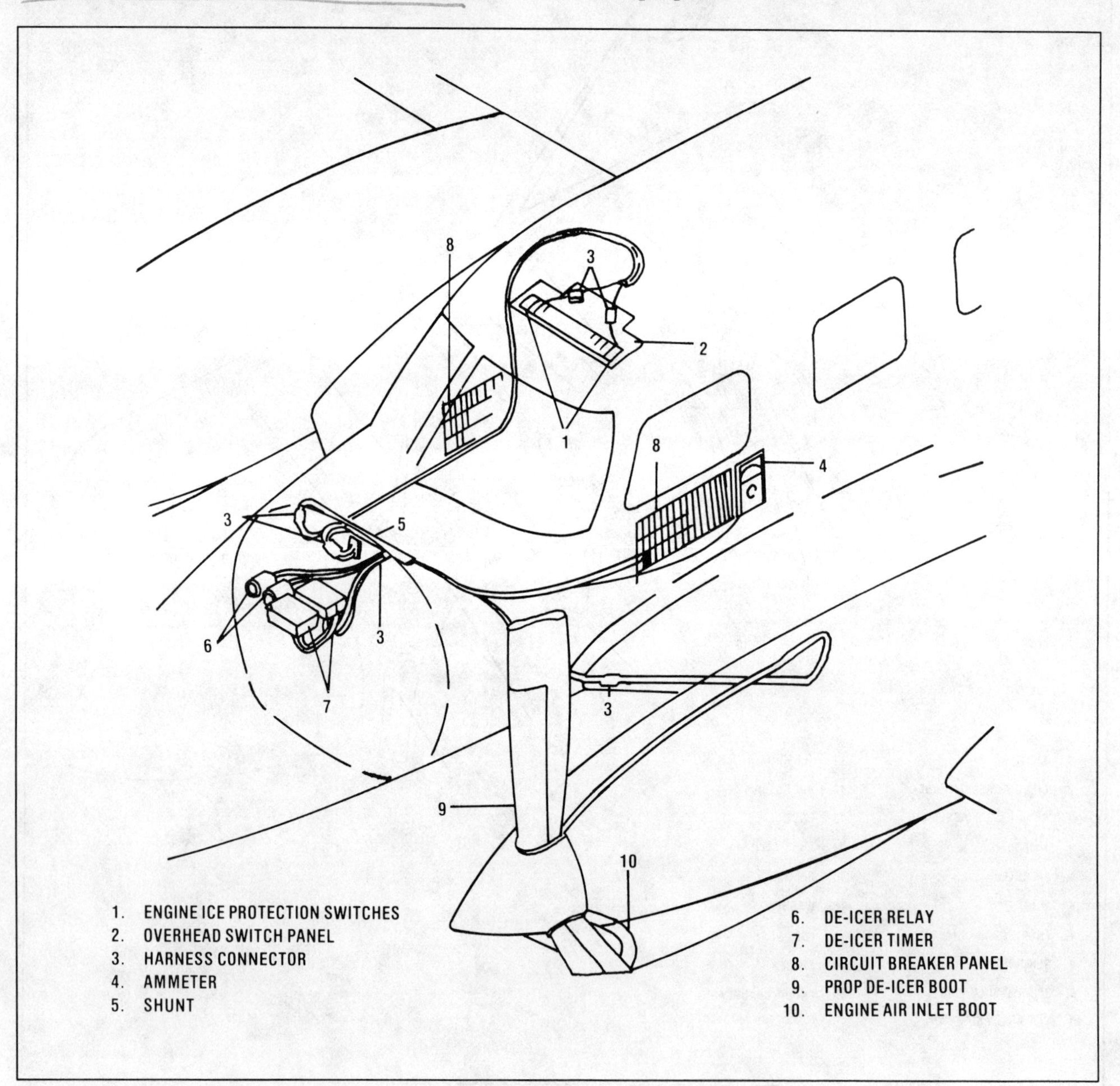

Figure 13-8. Electric de-icing system on a light turboprop.

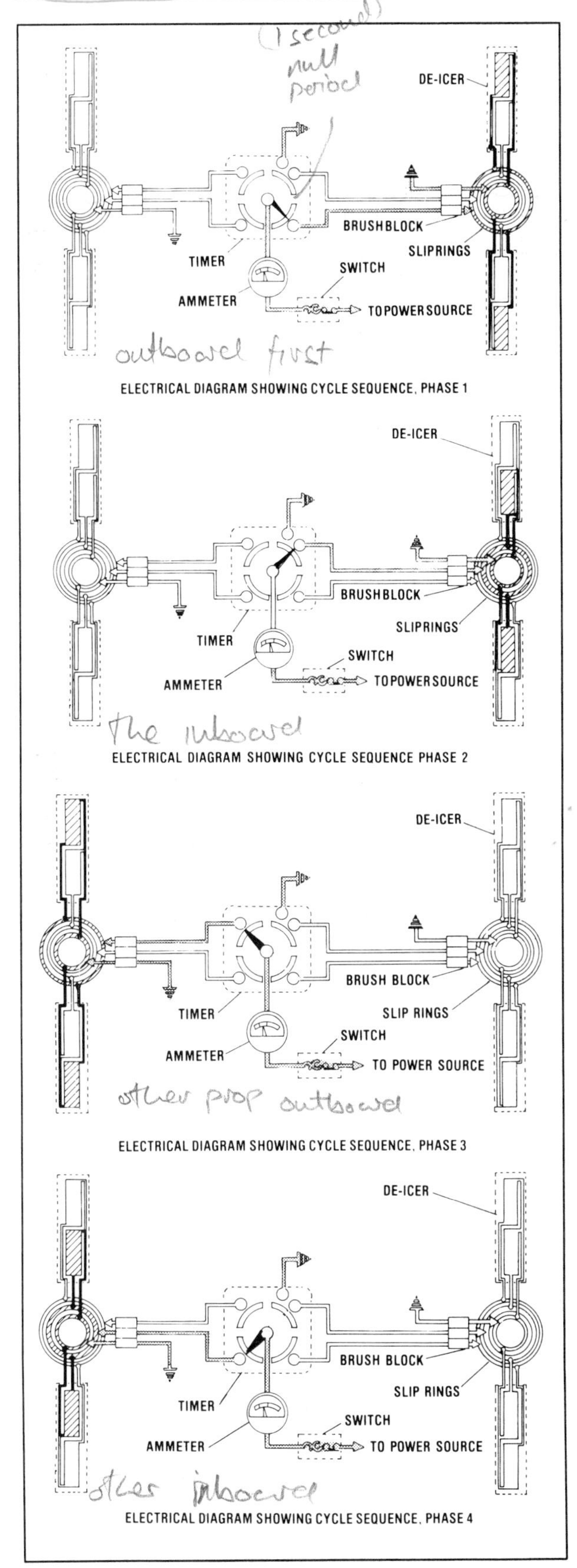

Figure 13-9. Cycling sequence of a light twin de-icing system using boots with two elements.

c. Inspection, Maintenance And Repair

Maintenance and inspection varies from system to system and the proper procedure for checking the operation of one system may damage or destroy the components of another system. The most sensitive components in the system are the blade boots. While one system may allow unlimited operation on the ground during inspection, another system may only allow two cycles of the system's heating sequence before it must be allowed to cool for 30 minutes or more because of the heat generated by the blade boots. For this reason, always consult the aircraft maintenance manual before performing any maintenance on the de-icing system.

The system should be inspected visually for any damage, wear, and for any dirt accumulation on the brush block and slip ring assemblies. Replace damaged and worn components and wires following the aircraft manufacturer's procedures. Clean the slip ring and brush block assemblies with an approved solvent and assure proper brush contact on the slip ring and sufficient brush length.

Electrical checks may be performed on the system components, both for resistance and for proper sequencing through voltage checks with the boots disconnected.

An operational check may be performed on some systems with the engines not operating using the following procedure. With an assistant in the cockpit to operate the controls and to time the sequence, turn on the aircraft master switch and the de-icing switch. Use your hands to determine which boot is heating. Follow the sequence of boot heating with your hands and have an assistant time the length of each sequence by observing the loadmeter indications. If all boots have a similar heat rise and the time that each boot is heated is in accordance with the service manual, the system is operating properly. If a system malfunction is noted, follow the aircraft manufacturer's instructions for troubleshooting the system. Note that when using battery power, the voltage applied to the timer is slightly less than full operating voltage, consequently, the system may take up to 10% longer to complete the de-icing cycle. Again, check the maintenance manual before performing this operational check.

De-icing system boots should be inspected for damage and electrical continuity. If necessary, the boots may be replaced by a mechanic following the aircraft manufacturer's instructions. Check closely for corrosion as mentioned for anti-icing shoes.

d. Troubleshooting

When troubleshooting the system, remember that the timer is the central component common to all propeller heating circuits. After the timer, all propellers have separate circuits and do not interfere with each other's operation. Therefore, if the system does not function properly for any propeller, the problem is with the timer or its electrical source.

If one propeller does not de-ice, the problem may be with the timer contact, the wires between the timer and the power relay, the brush block or brushes, or the slip ring. The boots are not likely to be a problem if no boot will heat. The problem is with a component that is common to all of the boots.

If one blade does not de-ice, the problem is either in the boot, the wires on the propeller, or the terminal on the propeller.

B. Simultaneous Propeller Control Systems

A simultaneous propeller control system provides a means by which the RPM setting of all engines can be changed by one control and/or a means of setting all engines at the same RPM.

1. Master Control System

A master control system allows the pilot to change the RPM of all engines an equal amount by the use of one lever, called the master control lever. The system also has the capability of setting all governors to their maximum RPM setting.

THE MASTER CONTROL IS CONNECTED TO A ROTARY CONTACT BY A CABLE. WHEN THE CONTROL IS MOVED, THE ROTARY CONTACT TOUCHES A BRUSH OF A FOLLOW UP ARM. THIS CONNECTION STARTS A DC MOTOR TURNING IN THE DIRECTION THE MASTER CONTROL HAS BEEN MOVED. THE FOLLOW UP ARM ROTATES WITH THE DC MOTOR UNTIL ITS BRUSH MOVES OFF THE ROTARY CONTACT.

PILOT'S MASTER CONTROL LEVER

BATTERY

ROTARY CONTACT

BRUSH

FOLLOW UP ARM GEARED TO DC MOTOR

INCREASE

DECREASE

RPM

PULLEYS AND CABLE

DC MOTOR

ELECTRICAL DIAGRAM SHOWING CYCLE SEQUENCE, PHASE 2

Figure 13-10. Hamilton-Standard master control system.

As this system requires the use of electric governor heads, it is most often used on transport aircraft.

a. System Components

The master control lever in the cockpit is used to adjust the RPM setting of all the governors and to set all governors at their maximum RPM setting. The lever moves through an arc in the same way that cable controls move and indicates the approximate RPM setting by its position in the arc of travel.

A rotary contact is connected to the master control lever through a cable and moves when the master control lever is moved. The contact completes an electrical path to a DC motor through a follow-up arm and causes the motor to drive the governor control boxes and change the setting of the governor head.

The reversible DC motor in the system is used to drive the governor commutator switches in the governor control boxes to reset the governor heads to a higher or lower RPM, depending on the direction in which the control lever is moved. When the DC motor rotates an amount proportional to the signal from the rotary contact, the follow-up arm moves off the rotary contact and the motor stops.

b. System Operation

The master control system is operational whenever the aircraft master control switch is turned on.

When the master control lever is moved forward, all engines will increase their RPM by the same amount. This is done by the rotary contact being moved by movement of the master control lever and directing electrical power to the DC motor,

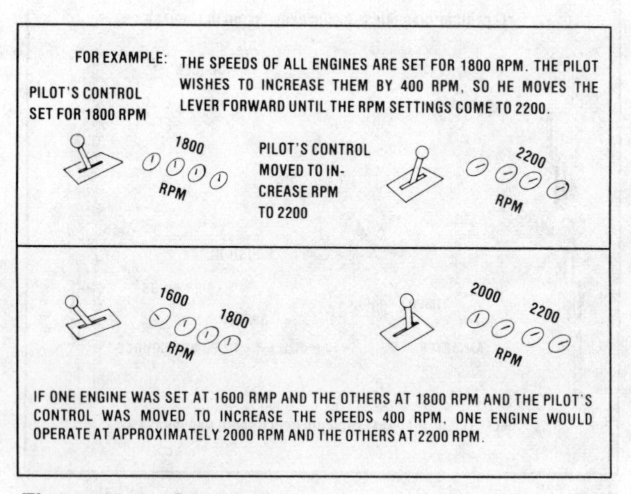

Figure 13-11. Governor response to movement of the master control lever.

rotating it in the proper direction. As the motor runs, the governors are reset to a higher RPM and the follow-up arm rotates until the arm runs off the contact area on the rotary contact. This gives a proportional response to the RPM increase. A large movement of the control lever will cause a large change in RPM before the follow-up arm runs off the contact area and a small control movement will cause a small change in RPM.

When the master control lever is moved aft, all governors will be reset for a lower RPM proportional to the control lever movement.

It is important to understand that all governors will change approximately the same RPM, but slight differences will always exist and the system will rarely be in synchronization by use of the master control system.

The master control system has the capability of setting all of the governors at their maximum RPM setting by moving the master control lever full forward. This will cause all of the governors to drive to their high RPM stop. This capability permits the pilot to set all engines easily at approximately the same RPM and eliminates the need to adjust the RPM of engines by use of the individual toggle switches. This is known as *calibrating* the system.

At any time the pilot has individual control of the propeller governors through the toggle switches.

c. Inspection, Maintenance And Repair

Maintenance of the master control system involves inspection, rigging, and cleaning of the system components.

The cable connecting the master control lever should be adjusted, lubricated, cleaned, and inspected in the same manner as for aircraft control cables. The rotary contacts should be inspected for arcing, pitting, and dirt, with the proper corrective maintenance. The DC motor should be inspected and maintained according to standard electrical practices.

d. Troubleshooting

If the system does not respond to movement of the control lever, one of the following may be the cause: no electrical power to the master control system; the control cable may be loose or broken; the rotary contact may be defective; the DC motor may not be operating.

If the master control lever causes an increase in RPM, but not a decrease, the decrease side of the follow-up arm may have an open circuit. If the RPM can be decreased, but not increased, the other contact on the follow-up arm may be open.

2. Synchronization System

The propeller ***synchronization system*** is used to set all governors at exactly the same RPM, thereby eliminating excess noise and vibration. A synchronization system may be used with governors having mechanical or electrical heads. Aircraft as small as the Cessna 310 may use a synchronization system.

The synchronization system is normally used for all flight operations except takeoff and landing. A master engine is used to establish the RPM to which all other engines (slave engines) will adjust.

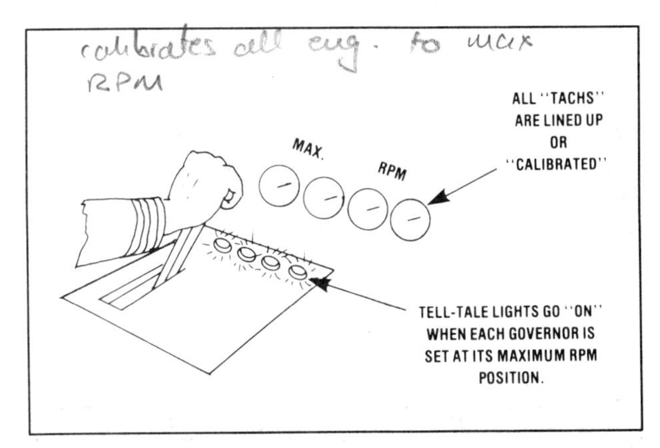

Figure 13-12. Calibrating the propellers with the master control lever.

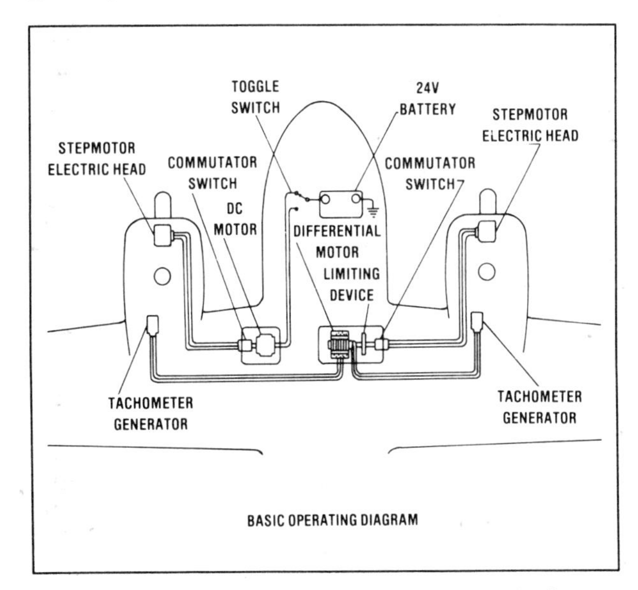

Figure 13-13. Diagram of the Hamilton-Standard synchronization system.

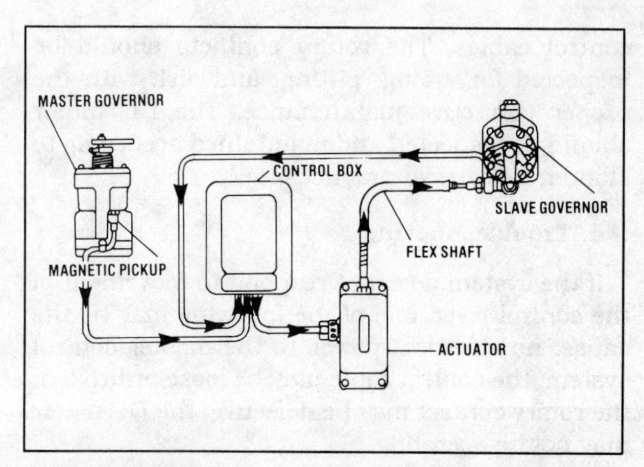

Figure 13-14. Woodward synchronization system for a light twin.

a. System Components

A ***tachometer-generator*** or a ***frequency generator*** used with each engine of a synchronization system generates a signal proportional to the RPM of the engine. The tach-generator is mounted on the rear accessory case of an engine. A frequency generator may be included in the governor construction.

A ***comparison unit*** is used to compare the RPM signal of the slave engines to the RPM signal of the master engine. If a tach-generator is used, the signal voltage is directed to a differential motor to compare the master engine RPM and the slave engine RPM. The engine which generates the higher voltage will determine the direction that the differential motor will rotate and adjust the governor setting of the slave engine. If a frequency generator is used, the engine signals are sent to an electronic unit which compares the frequencies and sends a correcting signal to the slave engine governor control mechanism.

The comparison unit has a limited range of operation and the slave engines must be within about 100 RPM of the master engine RPM for synchronization to occur.

A four-engine aircraft synchronization system may include a master engine selector switch which allows the pilot to select the master engine to be used (normally engine #2 or #3). This provides an alternate master engine if the engine being used as the master should become inoperative. A twin-engine aircraft uses the left engine as the master engine.

A resynchronization button is used in some systems to interrupt the synchronization system

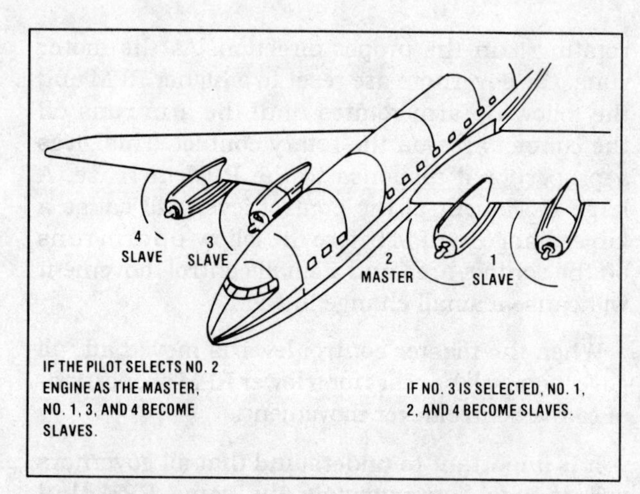

Figure 13-15. The master engine arrangement of a transport aircraft.

operation and allow the slave governor synchronization drive mechanisms to center, providing for full travel (100 RPM) toward the master engine RPM. This control is used if one or more slave engines are more than 100 RPM different from the master engine without the need to operate individual toggle switches.

b. System Operation

The synchronization system is used for all phases of flight except takeoff and landing. If the system were used for takeoff or landing, failure of the master engine would result in all the engines trying to follow the master engine and would cause a total system loss of power as the RPM of all engines decreased 100 RPM.

During normal operation, the slave engines are near the master engine RPM when the synchronization system is turned on. The signal comparison of the master engine and the slave engine signals through the comparison unit causes the slave engines' governors to adjust to the same RPM as the master engine.

If a master control system is incorporated with the synchronization system, the master control can be used at any time to adjust the RPM of all engines. As the master control lever is moved, the synchronization system is interrupted and the engines may go out of synchronization for a few seconds. When the lever stops moving, the system returns to synchronization.

The resynchronization button is used to recenter the synchronization system so that all engines can drive toward the master engine through their full range of travel (100 RPM).

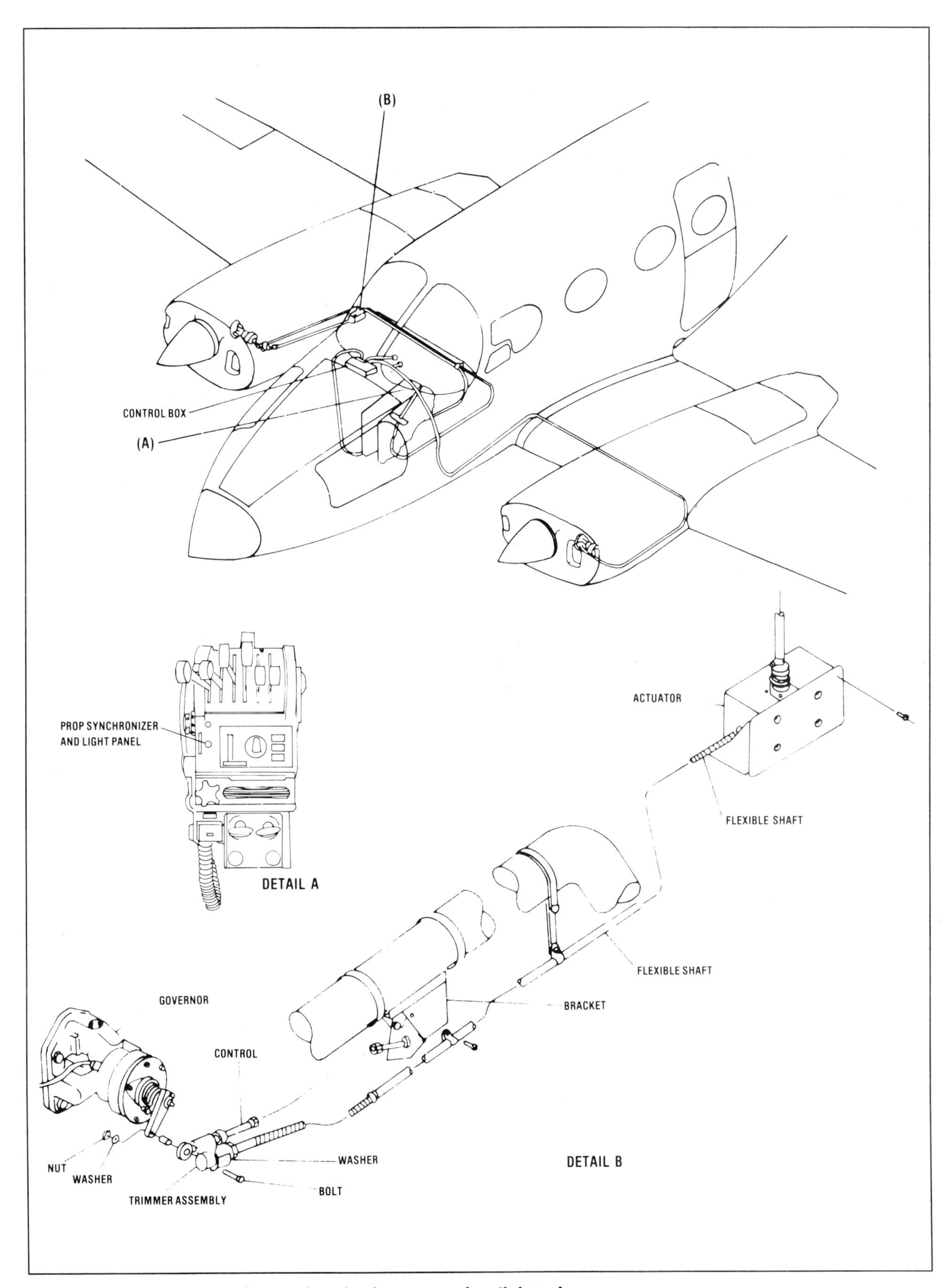

Figure 13-16. Installation of a synchronization system in a light twin.

c. Inspection, Maintenance And Repair

Maintenance of synchronization system involves assuring that the system is clean, lubricated, and electrically sound.

An operational check should be performed in a manner similar to the following: with the engines operating at a mid-range RPM, turn the synchronization system on and observe that the engine synchronize. Reduce the RPM of the master engine with the master engine's cockpit control and note that the slave engines follow the master engine for about 100 RPM. Resynchronize the system and reduce the RPM of each slave engine in small increments noting that the slave engine stays at the master engine RPM (or returns to the master engine RPM when the toggle switch is released) for a control movement equal to about 100 RPM. Outside of the 100 RPM range, the system should go out of synchronization.

If a resynchronization button is in the system, turn the system off and set the slave engines about 200 RPM different from the master engine RPM. Turn on the system and note that the slaves move toward the master RPM. Push the resynchronization button and the slave engines should move closer to the master engine RPM. Each time the button is pushed, the slaves should move 100 RPM toward the master engine RPM until all engines are in synchronization.

d. Troubleshooting

For troubleshooting of a system, refer to the service manual for the particular system and installation.

3. Synchrophasing System

Synchrophasing is a refinement of synchronization which allows the pilot to set the blades of the slave engines a number of degrees in rotation behind the blades of the master engine.

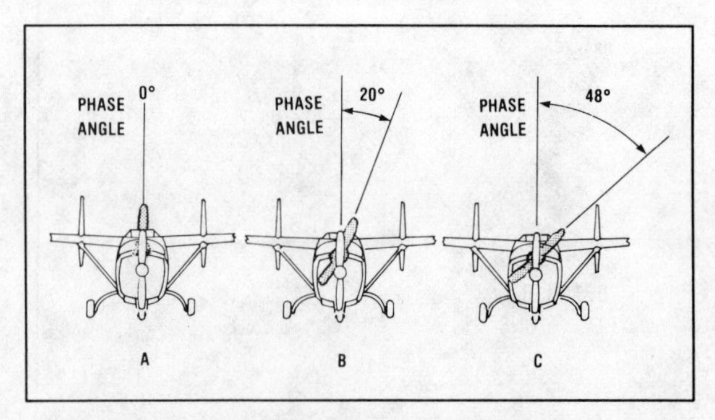

Figure 13-17. Synchrophasing sets the propellers of the aircraft at different angles and keeps them at the same RPM.

Synchrophasing is used to further reduce the noise created by the engines. The synchrophase angle can be varied by the pilot to adjust for different flight conditions and still achieve a minimum noise level.

A synchrophasing system may be used on aircraft as small as the Cessna 337.

a. System Components

A ***pulse generator*** is keyed to the same blade of each propeller (#1 blade for example) and the signal generated is used to determine if all #1 blades are in the same relative position at the same instant. The pulse generator serves the same function as the tach-generator in the synchronization system. By comparing when the signals from the slave pulse generators occur in relation to the master engine pulse, the mechanism will synchronize the phase relationship of the slaves to the master engine.

The synchrophaser electronic unit receives the signals from the pulse generators, compares them to the master engine signal, and sends a correcting signal to the governors. This adjusts the control of the slave engines to establish the phase angle selected by the pilot.

A propeller manual phase control in the cockpit allows the pilot to select the phase angle which will

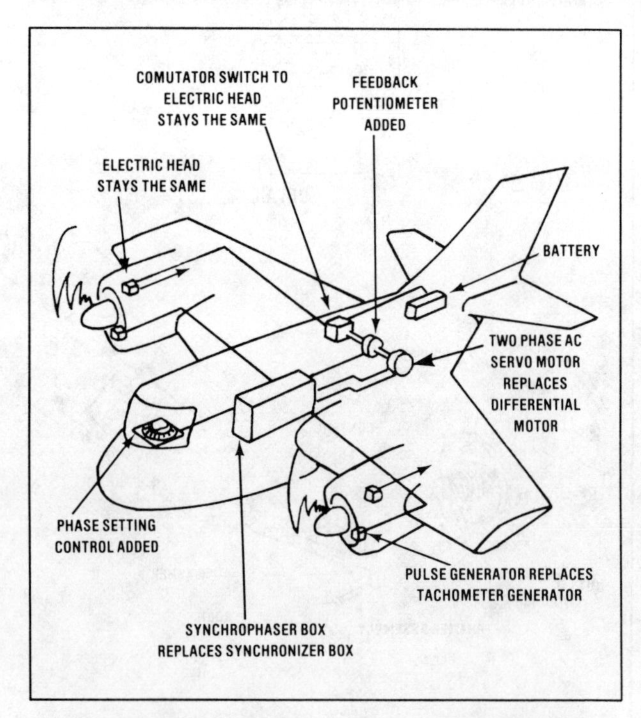

Figure 13-18. A comparison of the Hamilton-Standard synchronization system and synchrophasing system.

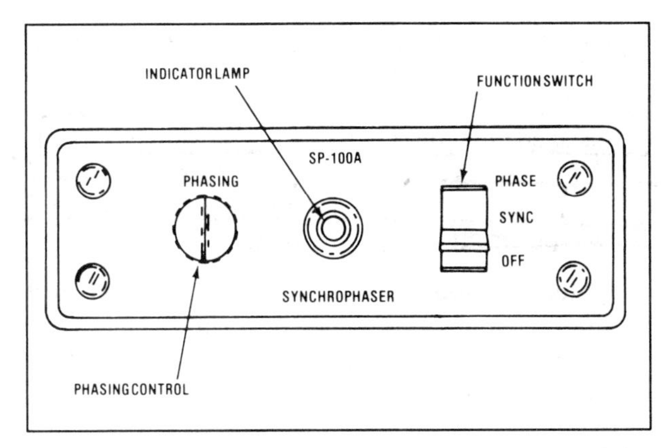

Figure 13-19. Synchrophasing control panel in a light twin.

give the minimum vibration. The control is a dial located with the propeller controls.

b. System Operation

When the engines are operating at nearly the same RPM, the system is turned on and the slaves will synchronize with the master engine. The electronic unit will adjust the governors to set the propellers at the phase angle set on the pilot's control dial. The pilot can vary the phase angle by rotating the control dial. This will cause the electronic unit to readjust the governors, setting the new phase angle.

c. Inspection, Maintenance And Repair

Please refer to the section on synchronizing for information.

d. Troubleshooting

For troubleshooting of a system, refer to the manual for the particular installation.

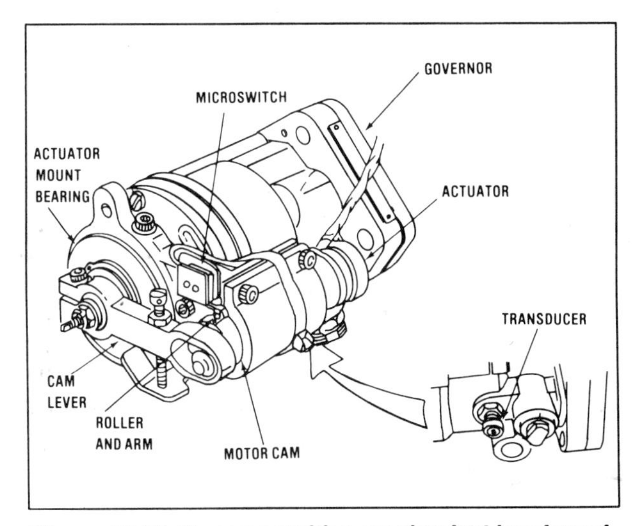

Figure 13-20. Governor with a mechanical head used with a light aircraft synchrophasing system.

C. Automatic Feathering System

An automatic feathering system is used on some transport category aircraft to feather a propeller automatically if the engine fails. The system is normally armed for takeoff and landing, but is turned off during cruising flight.

1. System components

(Please refer to Figure 13-21 for the following discussion.)

The system master switch is located on the pilot or flight engineer's console and is covered by a red guard. When the switch is turned on, a green indicator light appears to indicate that the system is armed.

A throttle switch is used to arm the circuit further by closing a microswitch when the throttle is advanced to a position of 45 to 75 percent of full throttle movement, depending on the aircraft. The circuit is open when the throttles are below this setting and the system will not autofeather.

A torque pressure switch is used to sense the power output of the engine and will close a contact whenever engine power drops below a specific power output. The amount of torque pressure loss required for the system to operate will vary with different aircraft due to engine size and aircraft design.

A time delay unit is used in the circuit to prevent autofeathering if only a momentary interruption in power occurs. The power loss must exceed one to two seconds for the system to autofeather. (This value varies with aircraft designs.)

A feather button is the unit ultimately activated by the system when an engine fails. A holding coil in the button is used to pull the feather button in

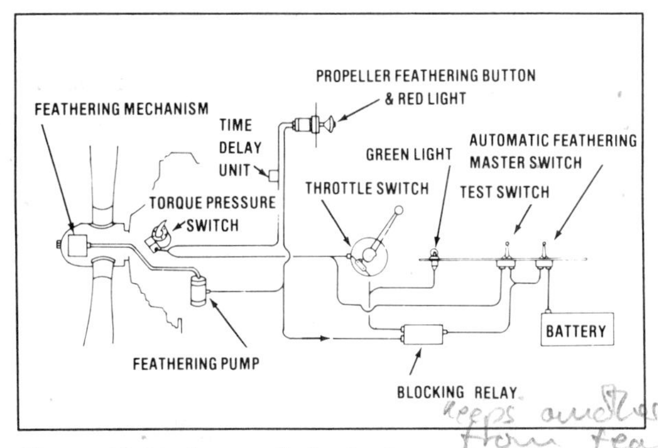

Figure 13-21. Automatic feathering system diagram for a transport aircraft.

and initiates the feathering operation in the same manner as if the pilot had pushed the button. When the button is actuated by the autofeather systems, a red light in the feather button illuminates to identify the feathered engine to the pilot. The button can also be operated in a normal manner by the pilot at any time.

A blocking relay is in the system to prevent more than one engine from autofeathering. This component may be located between the master switch and the throttle switch or may be incorporated in the feather button circuit. If one engine autofeathers, some systems can be reset to rearm the autofeather system in case another engine should fail. The pilot can feather any engine, at any time, by pushing the button regardless of whether or not a propeller has been autofeathered.

A test switch is used to bypass the blocking relay and throttle switch so that the system operation can be checked on the ground without developing high power settings.

2. System Operation

Before takeoff and landing, the system is armed by turning on the system master switch. As power is advanced for takeoff or during a missed landing approach, the throttle switch closes and the torque pressure switch is armed, but the torque pressure switch contacts are open. If a loss of engine power occurs, the torque pressure switch closes and, after the prescribed time interval, the time delay unit completes the circuit energizing the holding coil on the feather button. The button is pulled in initiating the feathering operation. At the same time, the blocking relay is actuated to break the circuit for the autofeather system on the other engines.

3. Inspection, Maintenance And Repair

An operational check of the system is the best way to check the system for operation and defects. Start the engines and arm the system with the autofeather system master switch. Advance the throttles to develop the required torque to arm the torque pressure switch. Hold the test switch in the position for the engine being checked and retard the throttle to idle. This should cause the torque pressure switch and the time delay relay to close and start the feathering operation by pulling in the feather button and illuminating the light in the button. Release the test switch and pull the button out to prevent the propeller from feathering. Note that with this check the blocking relay and throttle switch are not checked.

If a complete system check is desired and is not prohibited by the engine manufacturer, the same procedure is followed as above, but the test switch is not used and the throttle is advanced far enough to close the throttle switch. The mixture control can be pulled back to cause a loss in power and the propeller should autofeather.

System components should be inspected and maintained in accordance with the aircraft service manual. Units can be removed and replaced as necessary to correct system operation.

4. Troubleshooting

If the system indicator light does not illuminate when the system is armed, the bulb may be burned out, the system master switch may be open, or electrical power may not be getting to the system.

If the system operates properly during a ground test, but will not autofeather in flight, the throttle switch or blocking relay may be open or incorrectly adjusted.

If the system will not operate during a test or in flight, but the armed light is illuminated, the problem is most likely the torque pressure switch, the time delay unit, the feathering button holding coil, or the feathering system.

D. Pitch Lock

A pitch lock mechanism is used on some large transport aircraft to prevent an excessive engine overspeed if the governor should fail, preventing the blades decreasing their angle by centrifugal twisting moment. The pitch lock components are mounted in the dome assembly and low pitch stop-lever assembly in the Hamilton-Standard reversing Hydromatic® propellers. The Hamilton-Standard system is used for purposes of this discussion. Other systems work in a similar manner.

1. System Components

A flyweight-operated valve is located on the forward section of the stop-lever assembly, centered on the transfer tube, which is an oil passage through the assembly to the outboard side of the piston. The flyweight valve uses a spring to op pose the centrifugal force on the flyweights and sets the system so the valve will be moved onto the transfer tube blocking the oil passage when the system RPM reaches about 95% of the engine rated RPM. This prevents the engine oil pressure on the outboard side of the piston from returning to the engine oil pressure system and decreasing blade angle.

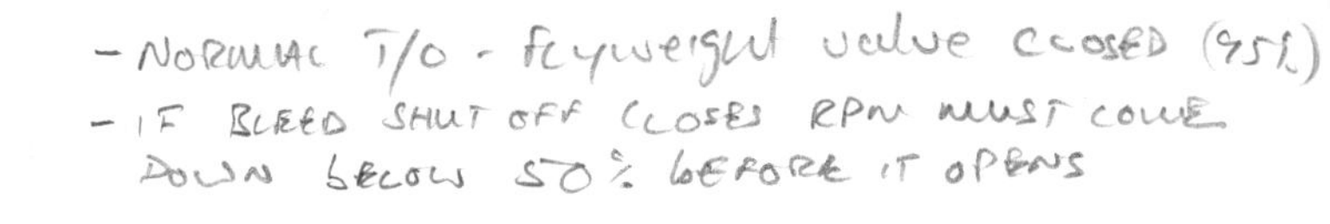

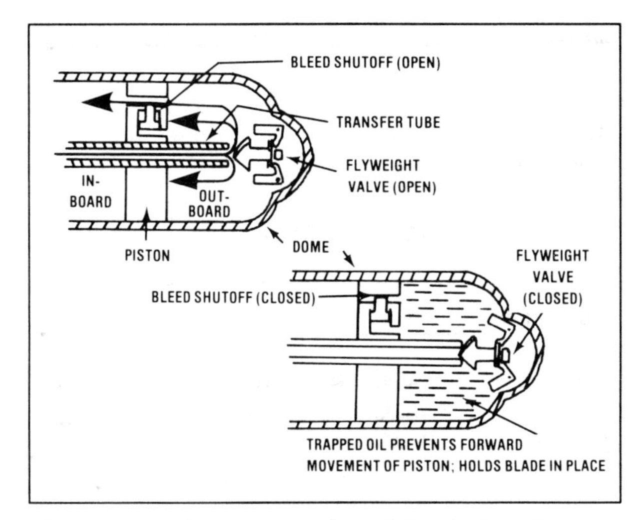

Figure 13-22. A representation of the pitch lock mechanism.

A bleed shutoff valve, consisting of a bleed valve and a spring, is located in each piston cam roller shaft. This used to close off the bleed passage between the inboard and outboard side of the piston if the system RPM exceeds the engine rated RPM by 10%.

2. System Operation

The system is automatic and is not controllable from the cockpit. As engine RPM increases, such as during a takeoff, the flyweight valve will close at 95% engine RPM preventing a rapid decrease in blade angle. Oil pressure from the governor will off-seat the valve and increase the blade angle as necessary to maintain rated RPM (100%). If the system overspeeds by 10%, the governor is no longer operating to control the RPM and may have failed. The bleed shutoff valve will now close and a hydraulic lock forming in the outboard portion of the propeller dome. This prevents oil from leaving the outboard side of the piston which would cause a decrease in blade angle and a greater overspeed. The bleed valve locks in position when it engages and will not retract until system RPM returns to some low value (about 50% RPM).

The propeller can be feathered at any time. It can also be reversed during the landing roll by the flyweight valve's being moved off its seat by the movement of the stop-lever assembly.

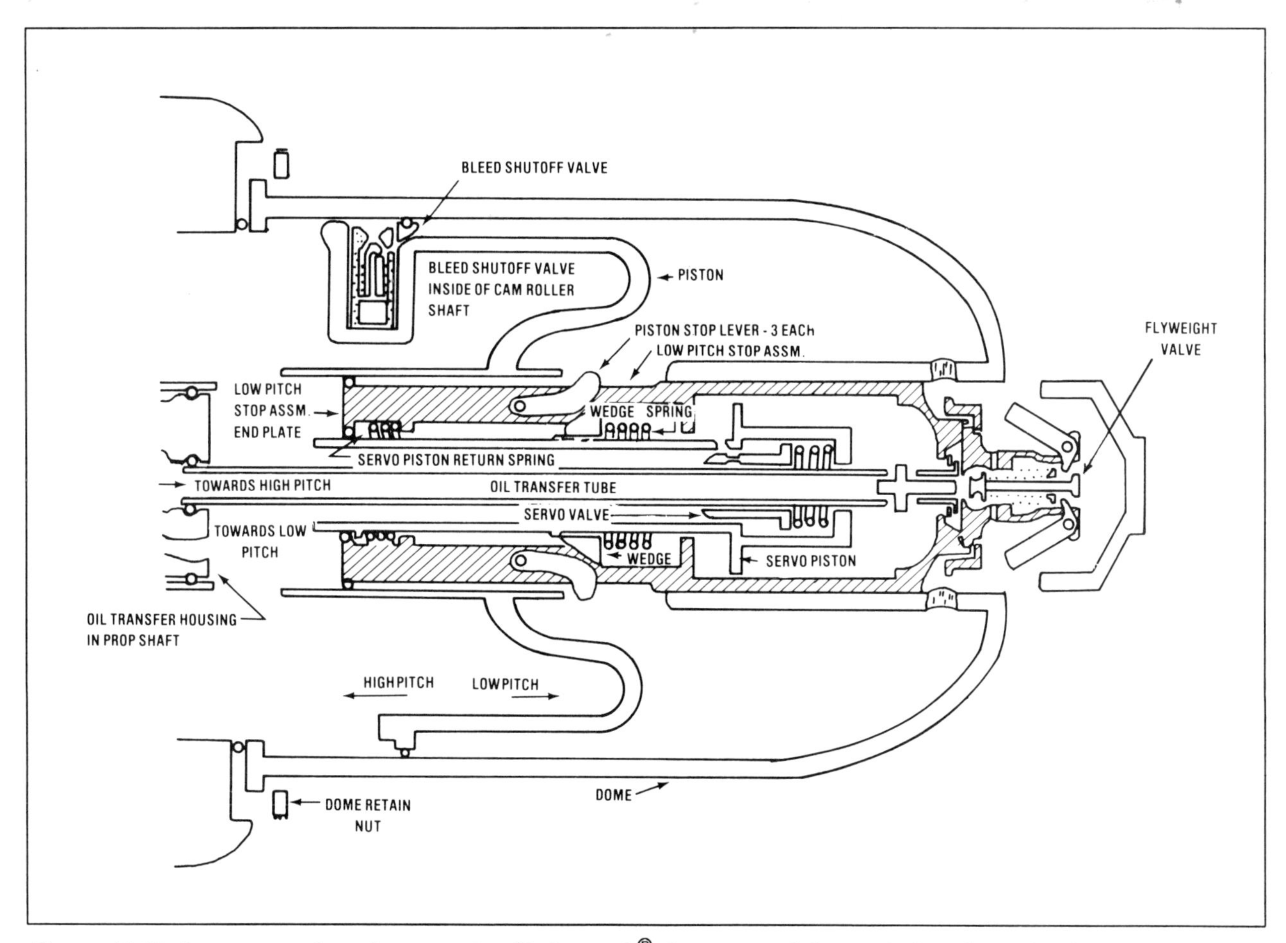

Figure 13-23. A cross section of a reversing Hydromatic® dome containing a pitch lock mechanism.

3. Inspection Maintenance And Repair

The pitch lock mechanism is an integral part of the propeller piston and the stop-lever assembly and no separate maintenance is required. The system is not accessible to the mechanic in the field and the propeller should be returned to an overhaul facility for correction of malfunctions.

4. Troubleshooting

If the pitch lock mechanism engages at too low an RPM or allows too high an overspeed, the valves may be defective or the unit may be set for the wrong RPM. (The same model propeller may fit different engines with different RPM limits.) If the overspeed tends to increase gradually, the fly-weight valve is closed, by the bleed shutoff valve may not be closing or the piston seals to the dome or the stop-lever assembly may be leaking.

Any malfunctioning of the system must be corrected by an overhaul facility.

E. Integral Oil Control Assembly

The ***Integral Oil Control Assembly*** (IOCA) is a self-contained propeller control unit mounted on the engine nose case directly behind the propeller. The IOCA contains all of the units necessary to control propeller operation. This system is used with reversing propeller installations on some large transport aircraft.

1. System Components

The IOCA components are contained in, or mounted on, a case which is bolted to the engine nose case.

Transfer sleeves in the center of the IOCA case are used to transfer oil to and from the rotating propeller from the stationary IOCA. The propeller shaft passes through the center of the IOCA.

The unit contains two internal pumps, a main pump and a scavenge pump. The main pump is located in the lower section of the case and takes oil from the IOCA sump, pressurizing it to a value necessary for propeller operation. The oil then moves to the governor where it is directed as necessary to and from the propeller to change the blade angles for constant-speed operation. The scavenge pump is used to return relieved system oil to the system sump.

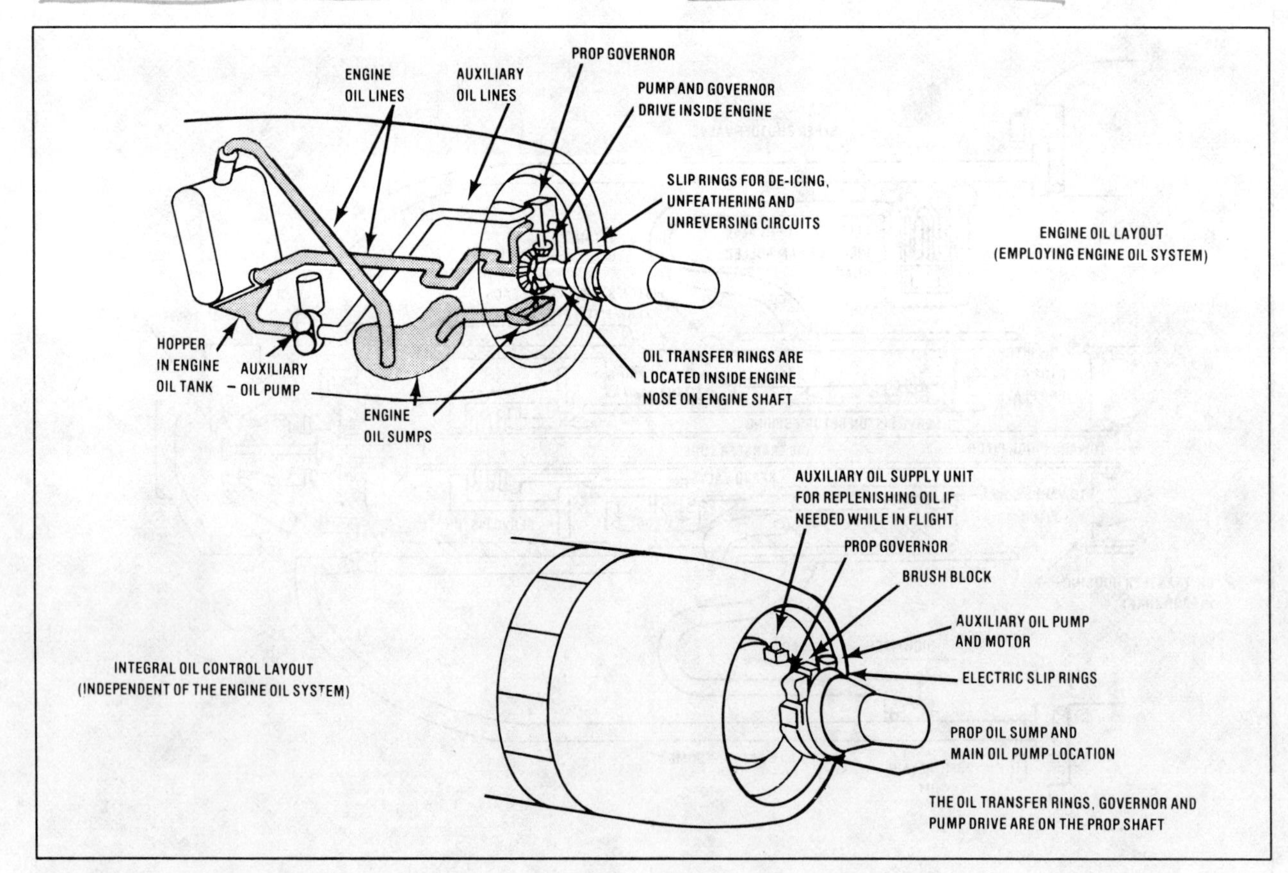

Figure 13-24. Comparison on the component location on a conventional propeller control system and the Integral Oil Control Assembly system.

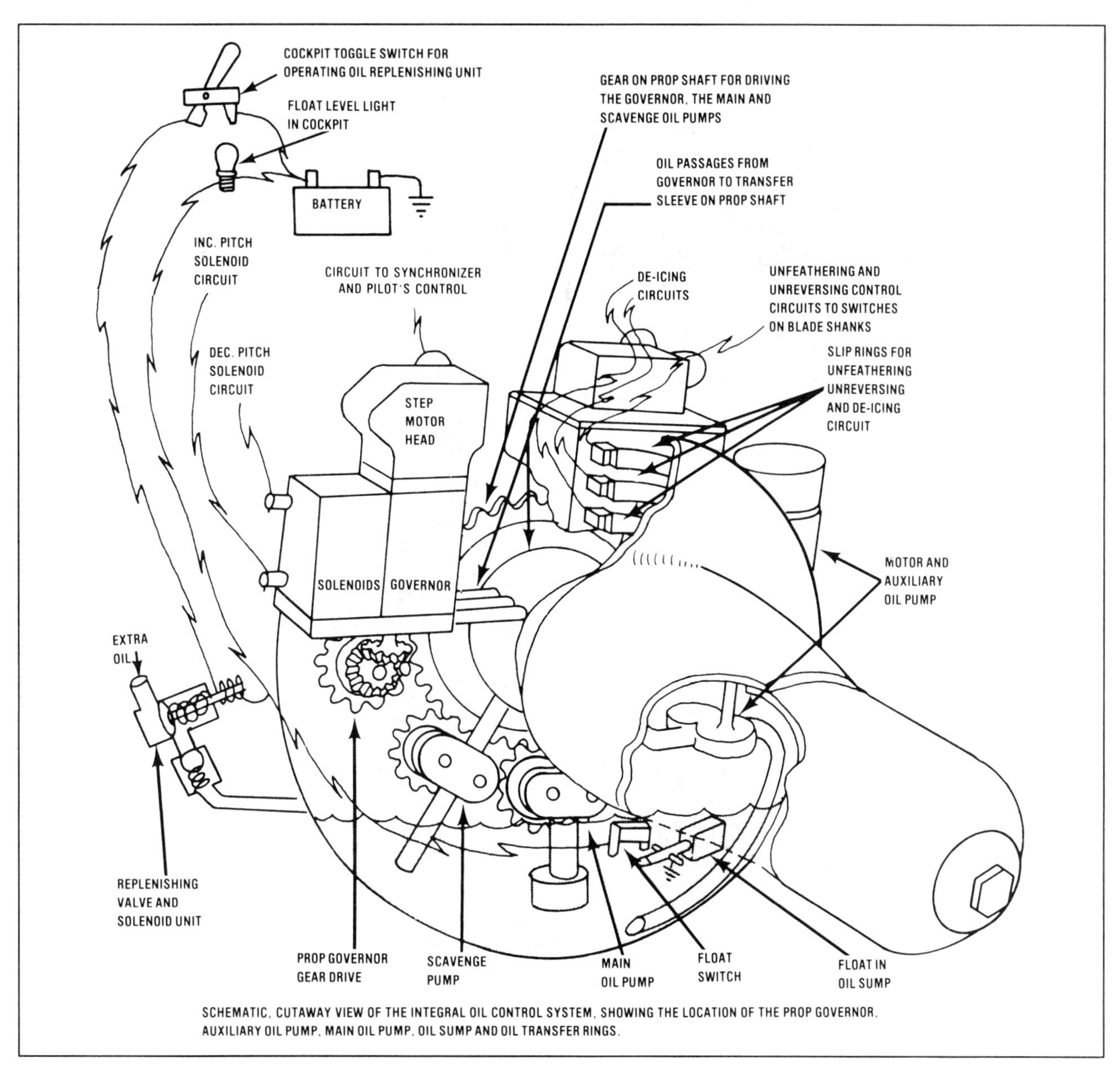

Figure 13-25. A front-view representation of an IOCA.

A high pressure relief valve is used to regulate system pressure as necessary to assure proper propeller operation.

A sump relief valve is used to maintain a slight air pressure in the system to reduce foaming of the oil during operation. Excess air pressure is vented overboard by the valve.

The main oil sump contains the system oil supply which is independent of the engine oil system. An auxiliary oil supply is located on the engine or in the nacelle to allow replenishment of the sump in flight. The oil used may be engine oil, hydraulic oil, or a special fluid, and may or may not contain a corrosion inhibitor.

External components mounted on the IOCA case include a governor, an auxiliary pump, and a brush block assembly.

2. System Operation

System operation for constant-speed, feathering and reversing operations is the same as for the Hydromatic® system discussed in other chapters of this book. A reserve supply of oil for feathering is obtained by placing the auxiliary oil pump pick-up below the level of the main pump pick-up.

If the system oil supply should become low in flight, an indicator light will come on in the cockpit. The pilot must then throw a switch to release the oil from the reverse supply tank.

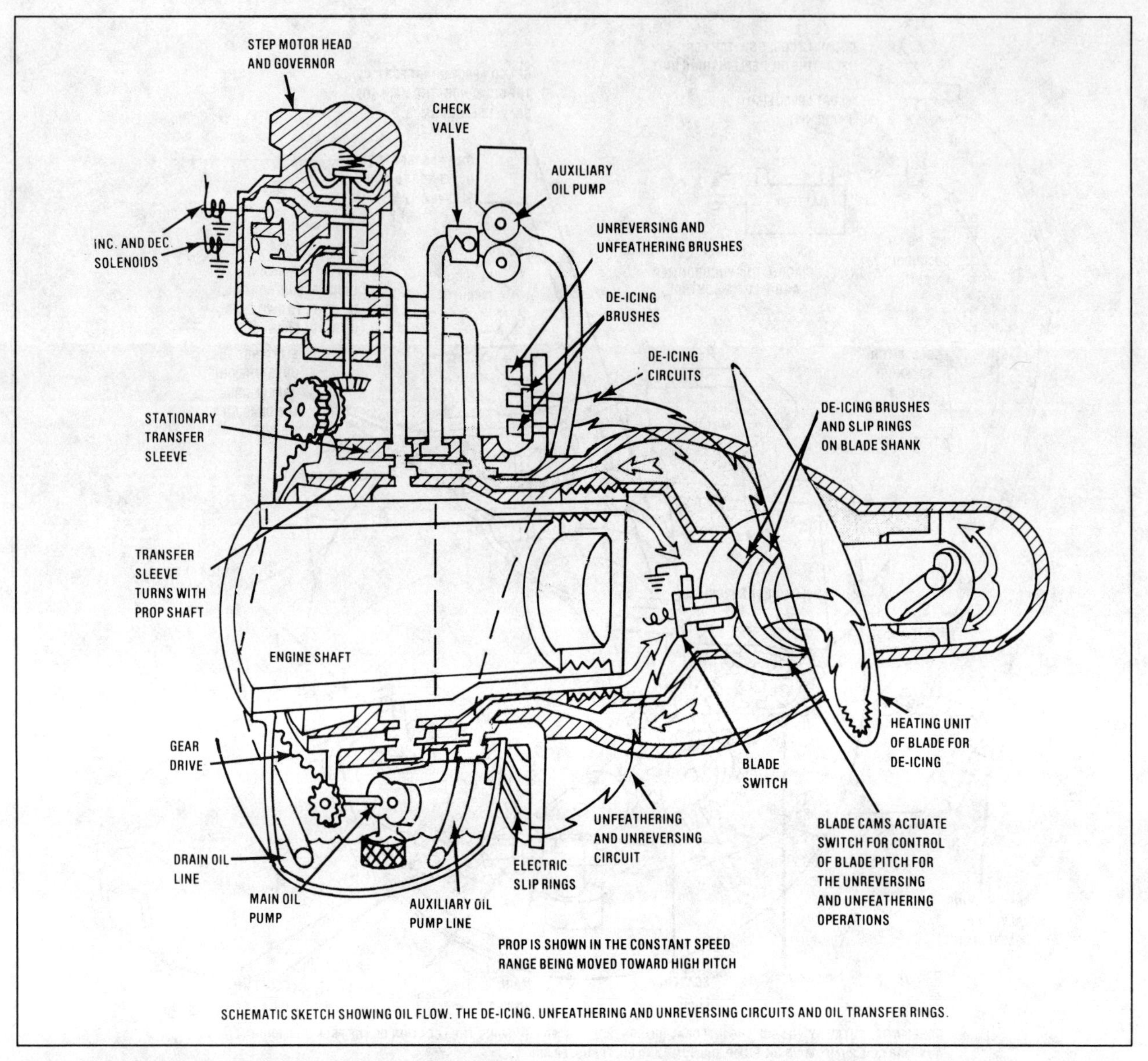

Figure 13-26. A side-view representation of an IOCA.

3. Installation

The IOCA is installed on the engine crankshaft, which drives the unit, and is bolted to the nose case. The propeller is then installed and the unit's cannon plugs and control cables are connected. Follow the instructions for the particular installation.

4. Inspection, Maintenance And Repair

The primary job of the mechanic is to install, service, adjust and replace components on the IOCA.

Servicing of the unit consists of replenishing the oil tanks, cleaning filters and screens, and keeping the brush blocks and propeller slip ring clean.

Adjustments include setting the governor RPM limits and aligning the brushes to the slip ring.

Some components on the IOCA can be removed and replaced. These include the governor, auxiliary pump and the brushes in the brush blocks assembly.

Maintenance involving disassembly of the IOCA unit is usually referred to an overhaul facility.

Refer to the appropriate maintenance manual for specific information concerning servicing of a particular installation.

5. Troubleshooting

Troubleshooting for the system is the same as for the individual systems in the IOCA, whether it be the reversing system, the ice elimination system, the synchrophasing system, etc.

QUESTIONS:

1. *Which ice elimination system prevents ice from forming?*
2. *What fluids may be used in an anti-icing system?*
3. *What is the purpose of the rheostat in the anti-icing system?*
4. *How are anti-icing feed shoes attached to the propeller blades?*
5. *What is the purpose of the Full De-ice Mode in a de-icing system?*
6. *What components are used to transfer the electrical power from the engine nose case to the propeller in a de-icing system?*
7. *Why is a null period included in the timing sequence in a de-icing timer?*
8. *What is the purpose of a master control system?*
9. *What is meant by calibrating the master control system?*
10. *What devices may be used to sense system RPM in a synchronization system?*
11. *Which engine is normally the master engine on a twin engine aircraft?*
12. *What is the purpose of the resynchronization button?*
13. *What is synchrophasing?*
14. *Which components of the automatic feathering system are bypassed during a normal ground operational check?*
15. *What is the purpose of the time delay unit in the auto-feather system?*
16. *What is the purpose of the pitch lock mechanism?*
17. *Which valve in the pitch lock mechanism closes during normal takeoff?*
18. *What routine maintenance can be performed on the pitch lock components?*
19. *What fluids may be used from propeller operation with an IOCA?*
20. *Which components on the IOCA may be replaced in the field?*

Glossary

This glossary of terms is provided to serve as a ready reference for the word with which you may not be familiar. These definitions may differ from those of standard dictionaries, but are in keeping with shop usage.

accumulator A device to aid in unfeathering a propeller.

aerodynamic twisting moment An operational force on a propeller which tends to increase the propeller blade angle.

angle of attack The angle between the chord line of a propeller blade section and the relative wind.

anti-icing system A system which prevents the formation of ice on propeller blades.

automatic propeller A propeller which changes blade angles in response to operational forces and is not controlled from the cockpit. Trade name: Aeromatic®.

back The curved side of a propeller airfoil section that can be seen while standing in front of the airplane.

blade One arm of a propeller from the hub to the tip.

blade angle The angle between the blade section chord line and the plane of rotation of the propeller.

blade index number The maximum blade angle on a Hamilton-Standard counterweight propeller.

blade paddle A tool used to turn the blades in the hub.

blade root The portion of a blade which is nearest the hub.

blade station A distance from the center of the propeller hub measured in inches.

boots Ice elimination components which are attached to the leading edge of propeller blades.

boss The center portion of a fixed-pitch propeller.

brush block The component of a de-icing and/or reversing system which is mounted on the engine nose case and holds the brushes which transfer electrical power to the slip ring.

centrifugal force The force on a propeller which tends to throw the blades out from the propeller center.

centrifugal twisting moment The force on a propeller which tends to decrease the propeller blade angle.

chord line The imaginary line which extends from the leading edge to the trailing edge of a blade airfoil section.

comparison unit The unit in a synchronization or synchrophasing system which compares the signals of the master engine and the slave engine and sends a signal to correct the slave engine RPM or blade phase angle.

cone The component used in a splined-shaft installation which centers the propeller on the crankshaft.

constant-speed system A system which uses a governor to adjust the propeller blade angle to maintain a selected RPM.

controllable-pitch propeller A propeller whose pitch can be changed in flight by the pilot's control lever or switch.

critical range The RPM range at which destructive harmonic vibrations exist.

de-icing system An ice elimination system which allows ice to form and then breaks it loose in cycles.

dome assembly The pitch-changing mechanism of a Hydromatic® propeller.

effective pitch The distance forward that an aircraft actually moves in one revolution of the propeller.

face The flat or thrust side of a propeller blade.

feather The rotation of the propeller blades to an angle of about 90 degrees which will eliminate the drag of a windmilling propeller.

fixed-pitch propeller A propeller, used on light aircraft, whose blade angles cannot be changed.

flanged shaft A crankshaft whose propeller mounting surface forms a flat plate 90 degrees to the shaft centerline.

frequency generator The engine RPM signal generator for some synchronization systems.

geometric pitch The theoretical distance that an aircraft will move forward in one revolution of the propeller.

governor The propeller control device in a constant-speed system.

go no-go gauge A gauge used to measure wear between the splines of a splined crankshaft.

ground-adjustable propeller A propeller which can be adjusted on the ground to change the blade angles.

hub The central portion of a propeller which is fitted to the engine crankshaft and carries the blades.

Hydromatic® A trade name for one type of Hamilton-Standard hydraulically operated propellers.

integral oil control assembly A self-contained propeller control unit used on some transport aircraft.

leading edge The forward edge of a propeller blade.

overhaul facility An FAA approved facility for major overhauls and repairs.

pitch The same as geometric pitch. Often used interchangeably with blade angle.

pitch distribution The twist in a propeller blade along its length.

pitch lock A mechanism used on some transports to prevent excessive overspeeding of the propeller if the governor fails.

plane of rotation The plane in which the propeller rotates, 90 degrees to the crankshaft centerline.

propeller A device for converting engine horsepower into usable thrust.

propeller disc The disc-shaped area in which the propeller rotates.

propeller repair station See overhaul facility.

propeller track The arc described by a propeller blade as the propeller rotates.

pulse generator The unit which generates an RPM and blade position signal in a synchrophasing system.

radial clearance The distance from the edge of the propeller disc to an object near the edge of the disc, perpendicular to the crankshaft centerline.

reversing Rotation of the propeller blades to a negative angle to produce a braking or reversing thrust.

safetying The installation of a safety device such as safety wire or a cotter pin.

selector valve Propeller control unit in a two-position propeller system.

shank The thickened portion of the blade near the center of the propeller.

shoe See boot.

shoulder The flanged area on the butt of a propeller blade which is used to retain the propeller blades in the hub.

slinger ring The fluid distribution unit on the rear of a propeller hub using an anti-icing system.

slip The difference between geometric pitch and effective pitch.

snap ring A component of a splined or tapered shaft installation which is used to aid in removal of the propeller.

spider The central component on many controllable-pitch propellers which mounts on the crankshaft and has arms on which the blades are installed.

splined shaft A cylindrical-shaped crankshaft extension which has splines on its surface to prevent propeller rotation on the shaft.

static RPM The maximum RPM that can be obtained at full throttle on the ground in a no-wind condition.

synchronization system A system which keeps all engines at the same RPM.

synchrophasing system A refined synchronization system which allows the pilot to adjust the blade relative position as they rotate.

tachometer-generator The RPM-sensing unit of some synchronization systems.

tapered shaft A crankshaft design whose propeller-mounting surface tapers to a smaller diameter and acts like a cone seating surface.

thrust bending force An operational force which tends to bend the propeller blades forward.

tip The portion of the blade farthest from the hub.

torque bending force An operational force which tends to bend the propeller blades in the direction opposite to the direction of rotation.

two-position propeller A propeller which can be changed between two blade angles in flight.

Answers to Study Questions

Chapter I

1. Convert engine horsepower to useful thrust.
2. The curved side of the airfoil section.
3. The angle between the airfoil section chord line and the plane of rotation.
4. Pitch distribution.
5. A propeller whose blade angle can be adjusted only on the ground.
6. Reduced landing roll and improved ground maneuverability.
7. Two.
8. 0.1 degree.
9. Mags *off*, throttle at *idle*, and mixture at *idle cutoff*.
10. Yes.

Chapter II

1. FAR 43.
2. Builder's name, model designation, serial number, type certificate number, and production certificate number.
3. A no-wind condition exists and the propeller rotation is not aided by the wind moving through the propeller.
4. Throttle.
5. Red.
6. Seven inches.
7. A reserve oil supply must be available only to the feathering pump.
8. Propeller repairman's certificate.
9. Major repair.
10. A maintenance release tag.

Chapter III

1. Tip.
2. The gradual change in blade angle from the root to the tip.
3. Decrease.
4. Two to four degrees.
5. Centrifugal force.
6. To reduce operational stresses.
7. Centrifugal twisting moment.
8. Six inches in from the tip.
9. A red arc on the tachometer.
10. The distance that the aircraft moves forward in one revolution of the propeller.

Chapter IV

1. Yellow birch.
2. Increase the structural strength of the tip.
3. Release moisture.
4. Cracks in the solder safety.
5. Can be repaired by an overhaul facility.
6. In a horizontal position in a cool, dry, dark area.
7. More efficient, better engine cooling, and require less maintenance.
8. To prevent a surface defect from developing into a crack.
9. Mild soap and water.
10. The blade shank.
11. Major repair.
12. 1C172 basic design; DM mounting and design changes; 75 inch diameter; 53 inch pitch.

Chapter V

1. Prevent corrosion and allow easy propeller removal.
2. So that the propeller will stop in a convenient position for hand propping.
3. Adapt the propeller for mounting on a tapered or splined crankshaft.
4. 70%.
5. Aid in removal of the propeller.
6. 20%.
7. Center the propeller on the crankshaft.
8. When the rear cone apex prevents the propeller hub from seating on the rear cone. Remove up 1/16-inch from the apex of the rear cone.
9. When the cone bottoms on the crankshaft splines before the propeller hub seats on the rear cone. Place a spacer of no more than 1/8-inch thickness behind the rear cone.
10. 1/16-inch.
11. Shims between the propeller and the hub or flange on wood propellers.

12. Clevis pin and cotter pin.
13. Uneven moisture distribution in the propeller.

Chapter VI

1. With grooves in the hub mating with shoulders on the blades.
2. Bolts.
3. The hub clamp rings or bolts and the propeller retaining nut.
4. White lead, red lead, or grease pencil.
5. Because of blade droop.
6. 0.1 degrees.
7. The blade retention area.
8. When the engine is overhauled.

Chapter VII

1. Centrifugal force and aerodynamic lift on the blade.
2. Airspeed and RPM.
3. Increase.
4. To prevent the propeller blades from being at different angles as the forces act to change the blade angles.
5. Two coats of clear nitrate dope applied over the defect.
6. Sluggish or erratic operation.
7. Fill the hub with oil.
8. Added.
9. Because the airflow over the propeller counterweights may affect propeller blade angle change.
10. Adjust the track with shims on a flanged installation or have the propeller overhauled.

Chapter VIII

1. A crank handle.
2. Through stops on the gear behind the propeller.
3. Binding or catching as the cockpit control is rotated.
4. No lubricant is used.

Chapter IX

1. 30.
2. Centrifugal force.
3. Rearward.
4. Prevent congealing of the oil in the piston, protect the piston from corrosion, prevent starvation of the engine bearings during engine start.
5. Piston.
6. 1 and 7.
7. 0.1 degrees.
8. Balancing.
9. Final balancing of the propeller with grease.
10. When a stop nut is positioned to set a blade angle three degrees or more from the minimum or maximum setting.

Chapter X

1. Governor.
2. To control the flow of oil to and from the propeller.
3. Flyweight assembly.
4. Decrease.
5. Decrease.
6. Increase.
7. Caused by oil or preservative in the governor and no action is necessary.
8. Raised side of the screen toward the governor.
9. 1/8-inch.
10. To set the governor at a cruise setting if the control cable from the cockpit should break.
11. Sluggish operation.
12. Rapid opening of the throttle.
13. Springs, centrifugal twisting force, and governor oil pressure.
14. To provide a constant dye penetrant inspection.
15. Once.
16. Between the engine crankshaft and the propeller hub.
17. The piston oil seal is leaking.
18. The Steel Hub propeller has its pitch-changing mechanism externally mounted.
19. Governor oil pressure and centrifugal twisting moment.
20. By rotating the blades in the blade clamps.
21. Remove one of the grease fittings in the blade clamp.
22. Loose blade clamps, defective seals, zerk fittings loose or leaking, over lubrication.

Chapter XI

1. Springs and centrifugal force on the counterweights.
2. Aid in unfeathering the propeller.
3. Ball and cylinder.
4. Place the aircraft in a shallow dive.
5. 100 psi.
6. Air pressure springs and centrifugal force on the counterweights.
7. Move the propeller control full aft.
8. Barrel assembly, dome assembly, and distributor valve.
9. Dome assembly.
10. To block the governor out of the system when feathering and unfeathering the propeller.
11. The pressure cutout switch.
12. Engine oil pressure.
13. Unfeathering.
14. Mid-range.
15. Feather.

Chapter XII

1. Decreased landing roll, reduced brake wear, increased ground maneuverability.
2. Insufficient engine cooling when reversing.
3. Throttle or power lever.
4. The landing gear must be extended and the aircraft weight must be on the gear.
5. Springs and centrifugal force on the counterweights.
6. Underspeed governor and propeller governor.
7. Propeller pitch control and propeller governor.
8. Ground operation.
9. Releases oil from the propeller.
10. RPM through the underspeed governor. RPM through the propeller governor.
11. PT6.
12. Propeller governor.
13. The pilot valve.
14. Fuel flow and propeller blade angle.
15. Fuel flow.
16. Set the low blade angle for constant-speed operation.
17. Reverse and feather.
18. Governor oil pump.
19. Prevent reversing before the aircraft weight is on the landing gear.
20. Unfeathering.

Chapter XIII

1. Anti-icing
2. Isopropyl alcohol or phosphate compound.
3. To control the rate of fluid flow.
4. With adhesives.
5. To de-ice all of the propellers at the same time.
6. A brush block and a slip ring.
7. So that only one propeller will de-ice at any one time.
8. To allow adjustment of all engine RPM settings with one control lever.
9. Setting all governors at their high RPM limits by moving the master control lever full forward.
10. A tachometer-generator or a frequency generator.
11. Left engine.
12. To allow all governors to drive toward the master engine RPM through full range each time the button is pushed.
13. Synchronizing and setting the blades at a specific angle in rotation behind the master engine.
14. Blocking relay and throttle switch.
15. To prevent autofeathering of a propeller during momentary interruptions in engine power.
16. Prevent excessive overspeeding of the engine.
17. Flyweight valve.
18. None.
19. Engine oil, hydraulic fluid, or special fluids.
20. Auxiliary pump, governor, and brushes.

Final Examination

Aircraft Propellers and Controls

Student ______________________

Grade ______________________

Place a circle around the letter for the correct answer to each of the following questions.

1. The constant-speed control unit is also called a:
   A. Accumulator.
   B. Governor.
   C. Selector valve.
   D. Propeller pitch control.

2. Which FAR lists the minimum inspections required during an annual inspection?
   A. 23.
   B. 25.
   C. 43.
   D. 65.

3. Which of the following is not a condition necessary to check the static RPM of an installation?
   A. Aircraft in a level attitude.
   B. Maximum allowable manifold pressure.
   C. No-wind condition.
   D. The aircraft is stationary.

4. What is the minimum water clearance of the propellers on a seaplane?
   A. 7 inches.
   B. 9 inches.
   C. 18 inches.
   D. 24 inches.

5. Who may supervise the major repair of a governor?
   A. A powerplant mechanic.
   B. An A&P mechanic with an Inspection Authorization.
   C. A propeller repairman.
   D. A powerplant repairman.

6. An aircraft is equipped with a fixed-pitch propeller. As airspeed increases:
   A. The blade angle increases.
   B. The blade angle decreases.
   C. The blade angle of attack increases.
   D. The blade angle of attack decreases.

7. Which force tends to decrease propeller blade angle?
   A. Centrifugal twisting moment.
   B. Aerodynamic twisting moment.
   C. Torque bending force.
   D. Thrust bending force.

8. A horizontal imbalance in a wood propeller may be corrected by:
   A. Solder on the boss.
   B. Solder on the tip.
   C. A plate on the boss.
   D. A plate on the tip.

9. The recommended maximum depth of repair on the trailing edge of an aluminum propeller blade is:
   A. 1/16-inch.
   B. 1/8-inch.
   C. 3/16-inch.
   D. 1/4-inch.

10. Which type of installation does not require the use of Prussian Blue to check for seating?
    A. Flanged shaft.
    B. Tapered shaft.
    C. Splined shaft.
    D. All require a Prussian Blue check.

11. Which force is used to increase the propeller blade angle of a Hamilton-Standard two-position propeller?
    A. Engine oil pressure.
    B. Springs.
    C. Centrifugal force.
    D. Governor oil pressure.

12. Which blade angle is used for takeoff in a controllable-pitch propeller?
    A. Reverse.
    B. Low blade angle.
    C. High blade angle.
    D. Feather.

13. What opposes the flyweights in a governor?
   A. Pilot valve.
   B. Balance spring.
   C. Transfer valve.
   D. Speeder spring.

14. Which cockpit control is used to make large changes in manifold pressure?
   A. Throttle.
   B. Propeller control.
   C. Mixture control.
   D. None of the above.

15. If an aircraft is equipped with a constant-speed system, which of the following will cause a decrease in propeller blade angle?
   A. Moving the propeller control aft.
   B. Moving the throttle forward.
   C. Placing the aircraft in a climb.
   D. All of the above.

16. What is an acceptable amount of cushion in a governor control lever?
   A. 1/8-inch.
   B. 1/4-inch.
   C. 3/8-inch.
   D. 1/2-inch.

17. What is the purpose of the high-pressure transfer valve in a feathering Hydromatic® system.
   A. Terminate the feathering operation.
   B. Block the governor out of the system.
   C. Shift oil passages to the propeller dome to allow unfeathering.
   D. Initiate the feathering operation.

18. Which turboprop control lever is used to control propeller blade angle in the Beta Mode?
   A. Speed lever.
   B. Condition lever.
   C. Propeller control lever.
   D. Power lever.

19. Which of the following switches is not used with a de-icing system?
   A. Rheostat.
   B. Full de-ice mode.
   C. Cycle speed selector.
   D. On-off switch.

20. A synchronization system is not used during:
   A. Takeoff.
   B. Climb.
   C. Cruise.
   D. Descent.

Answers to Final Examination Aircraft Propellers and Controls

1. B	11. C
2. C	12. B
3. A	13. D
4. C	14. A
5. C	15. C
6. D	16. A
7. A	17. B
8. B	18. D
9. B	19. A
10. A	20. A